Compendium of Soybean Diseases
THIRD EDITION

Edited by
J. B. Sinclair and P. A. Backman

In cooperation with

The Department of Plant Pathology
University of Illinois at Urbana-Champaign

The Department of Plant Pathology
Auburn University, Auburn, Alabama

APS PRESS
The American Phytopathological Society

Financial Sponsors

American Soybean Association
Southern Soybean Disease Workers

Front cover photograph by Nancy Kokalis, Auburn University
Back cover photograph by W. O. Scott, University of Illinois at Urbana-Champaign

Library of Congress Catalog Card Number: 88-83464
International Standard Book Number: 0-89054-093-4

Printed in the United States of America on acid-free paper

The American Phytopathological Society
3340 Pilot Knob Road
St. Paul, Minnesota 55121-2097, USA

Preface

The third edition of the *Compendium of Soybean Diseases* builds on the previous editions to provide a thorough, authoritative, and practical reference for plant pathologists and other agricultural researchers working with soybean diseases in the field, in diagnostic laboratories, or in plant clinics. It is designed for those training in soybean pathology and will assist agricultural workers in areas where soybeans are a recently introduced crop. This revised edition should prove even more helpful than the first two editions to crop disease consultants and their scouts, advisors in government departments of agriculture and regulatory agencies, agribusiness representatives, county agents, area crop specialists, educators, students, growers, and researchers the world over. This edition is international in scope.

Each section of the compendium has been revised and updated with the aid of agricultural scientists from around the world. The names of the persons who prepared the information on individual diseases appear at the ends of the sections. Sections without such acknowledgments were prepared by the editors. A list of contributors follows the preface.

Many new illustrations and several new sections have been added, including those on sudden death syndrome, red leaf blotch, *Bradyrhizobium*, and mycorrhizal fungi. The section on stem canker includes studies on the disease as it occurs in the southern United States.

The lists of references have been revised, to include a blend of contemporary and classic scientific papers, books, monographs, and proceedings of workshops and conferences. The first two editions of the compendium should be retained for their references. (For a review of earlier literature, consult J. B. Sinclair and O. D. Dhingra, 1975, *An Annotated Bibliography of Soybean Diseases, 1882–1974*, INTSOY Ser. 7, College of Agriculture, University of Illinois at Urbana-Champaign.)

The descriptions of diseases are arranged in sections according to general causal agents: diseases caused by bacteria and mycoplasmalike organisms, fungi, viruses, and nematodes; diseases caused by abiotic agents; and diseases of unknown or uncertain etiology. Because of the economic importance of factors affecting soybeans grown for seed, industrial use, food, and animal feed, a section on soybean seed pathology is also included. The most authoritative scientific names are used for pathogens, hosts, and vectors, along with the names of authors; synonyms for causal agents are given where appropriate.

Suggested control measures given at the ends of the descriptions are purposely general, stressing principles and cultural practices that should not become obsolete and are applicable in most soybean-growing areas of the world. A comprehensive discussion of strategies for controlling soybean disease is presented as a separate section. For current information on chemical control and cultural practices, readers should consult recent literature, competent area specialists, extension or advisory plant pathologists, or informed agricultural suppliers. (For information on resistance to some diseases prior to 1980, consult O. Tisselli, J. B. Sinclair, and T. Hymowitz, 1980, *Sources of Resistance to Selected Fungal, Bacterial, Viral and Nematode Diseases of Soybeans*, INTSOY Ser. 18, College of Agriculture, University of Illinois at Urbana-Champaign.)

A computer program entitled "Soybean Disease Diagnosis" has been developed at the University of Illinois at Urbana-Champaign (see R. S. Michalski, J. H. Davis, V. S. Bisht, and J. B. Sinclair, 1983, A computer-based advisory system for diagnosing soybean diseases in Illinois, Plant Dis. 67:459-463). The program is designed for use by remote terminals or personal computers on farms, at county agent or crop consultant offices, and in diagnostic clinics, to diagnose 15 major soybean diseases. For further information, contact the Illinet Office, University of Illinois at Urbana-Champaign, 123 Mumford Hall, 1301 West Gregory Drive, Urbana, IL 61801 USA. This compendium could be used as a companion piece to the computer program for more detailed information.

Further revisions of this book are planned, to keep the information up to date and to improve the format, the amount and level of detail, the quality of illustrations, and other items as needed. Comments regarding the general usefulness of this edition or omissions from it are welcome. Numerous suggestions made by readers and users of the previous editions are incorporated here. With your suggestions, future editions can be even more useful.

The personnel and facilities of the Department of Plant Pathology at Auburn University and the Department of Plant Pathology at the University of Illinois at Urbana-Champaign were used in the preparation of this edition, with the support of B. J. Jacobsen and R. E. Ford, respectively. Special thanks are given to Kim J. Morrison, of the University of Illinois at Urbana-Champaign, and to Tammy Forbus and Carole Backman, of Auburn University, for aid in preparing the manuscript. The invaluable assistance of Larry Dalrymple, of Auburn University, for general editing as well as for editing botanical nomenclature is gratefully acknowledged.

The editors wish to thank those who were willing to review and revise the material on disease illustrations and assisted in the production of this edition in numerous ways. Nancy M. Kokalis, of Auburn University, provided the scanning electron micrographs.

Authors

N. Acosta V.
Department of Crop Protection
University of Puerto Rico–Mayaguez

P. A. Backman
Department of Plant Pathology
Auburn University
Auburn, AL

G. T. Berggren, Jr.
Department of Plant Pathology and
  Crop Physiology
Louisiana State University
Baton Rouge, LA

K. J. Cavanaugh
Cornell University–IPM Support
  Group
Auburn, NY

W. S. Curran
Department of Agronomy
University of Illinois at
  Urbana-Champaign

L. E. Datnoff
Institute of Food and Agricultural
  Sciences
Agricultural Research and Education
  Center
University of Florida
Belle Glade, FL

J. W. Demski
University of Georgia
Georgia Agricultural Experiment
  Station
Experiment, GA

J. M. Dunleavy
U.S. Department of Agriculture
Agricultural Research Service
Department of Plant Pathology
Iowa State University
Ames, IA

D. I. Edwards
U.S. Department of Agriculture
Agricultural Research Service
Department of Plant Pathology
University of Illinois at
  Urbana-Champaign

B. A. Fulling
Department of Plant Pathology
University of Illinois at
  Urbana-Champaign

W. S. Gazaway
Department of Plant Pathology
Alabama Cooperative Extension
  Service
Auburn University
Auburn, AL

D. A. Glawe
Department of Plant Pathology
University of Illinois at
  Urbana-Champaign

C. R. Grau
Department of Plant Pathology
University of Wisconsin
Madison, WI

L. E. Gray
U.S. Department of Agriculture
Agricultural Research Service
Department of Plant Pathology
University of Illinois at
  Urbana-Champaign

K. C. Hadden
Department of Plant Pathology and
  Crop Physiology
Louisiana State University
Baton Rouge, LA

A. K. Hagan
Department of Plant Pathology
Alabama Cooperative Extension
  Service
Auburn University
Auburn, AL

G. L. Hartman
International Institute of Tropical
  Agriculture
Ibadan, Nigeria

D. R. Hattermann
Department of Plant Pathology
University of Tennessee
Knoxville, TN

A. S. Heagle
U.S. Department of Agriculture
Agricultural Research Service
North Carolina State University
Raleigh, NC

D. E. Hershman
Research and Educational Center
University of Kentucky
Princeton, KY

A. D. Hewings
U.S. Department of Agriculture
Agricultural Research Service
Department of Plant Pathology
University of Illinois at
  Urbana-Champaign

M. C. Hirrel
Southeast Research and Extension
University of Arkansas
Monticello, AR

T. Hymowitz
Department of Agronomy
University of Illinois at
  Urbana-Champaign

B. J. Jacobsen
Department of Plant Pathology
Auburn University
Auburn, AL

B. W. Kennedy
Department of Plant Pathology
University of Minnesota
St. Paul, MN

R. A. Kinloch
Agricultural Research Center
University of Florida
Jay, FL

D. E. Kuhlman
Agricultural Entomology
State Natural History Survey
Urbana, IL

C. W. Kuhn
Department of Plant Pathology
University of Georgia
Athens, GA

M. M. Kulik
U.S. Department of Agriculture
Agricultural Research Service
Beltsville, MD

S. A. Lewis
Department of Plant Pathology and
  Physiology
Clemson University
Clemson, SC

S. M. Lim
U.S. Department of Agriculture
Agricultural Research Service
Department of Plant Pathology
University of Illinois at
  Urbana-Champaign

Z. Liu
Department of Plant Pathology
University of Illinois at
  Urbana-Champaign

J. L. Lockwood
Department of Botany and Plant
  Pathology
Michigan State University
East Lansing, MI

J. B. Manandhar
Plant Pathology Division
Khumaltar, Lalipur, Nepal

E. C. McGawley
Department of Plant Pathology and
  Crop Physiology
Louisiana State University
Baton Rouge, LA

D. C. McGee
Seed Science Center
Iowa State University
Ames, IA

M. D. McGlamery
Department of Agronomy
University of Illinois at
  Urbana-Champaign

R. A. Meronuck
Department of Plant Pathology
University of Minnesota
St. Paul, MN

J. S. Mignucci
Department of Crop Protection
University of Puerto Rico–Mayaguez

G. Morgan-Jones
Department of Plant Pathology
Auburn University
Auburn, AL

G. R. Noel
U.S. Department of Agriculture
Agricultural Research Service
Department of Plant Pathology
University of Illinois at
  Urbana-Champaign

D. V. Phillips
Department of Plant Pathology
Georgia Agricultural Experiment
  Station
Experiment, GA

R. D. Riggs
Department of Plant Pathology
University of Arkansas
Fayetteville, AR

R. Rodríguez-Kábana
Department of Plant Pathology
Auburn University
Auburn, AL

J. C. Rupe
Department of Plant Pathology
University of Arkansas
Fayetteville, AR

S. C. Schmidt
Department of Agricultural Economics
University of Illinois at
  Urbana-Champaign

D. P. Schmitt
Department of Plant Pathology
North Carolina State University
Raleigh, NC

A. F. Schmitthenner
Department of Plant Pathology
Ohio State University
Wooster, OH

J. B. Sinclair
Department of Plant Pathology
University of Illinois at
  Urbana-Champaign

C. Sittigul
Department of Plant Pathology
Faculty of Agriculture
Chiangmai University
Chiangmai, Thailand

J. P. Snow
Department of Plant Pathology and
  Crop Physiology
Louisiana State University
Baton Rouge, LA

K. L. Steffey
Agricultural Entomology
State Natural History Survey
Urbana, IL

N. Suwarnarat
Department of Plant Pathology
Faculty of Agriculture
Chiangmai University
Chiangmai, Thailand

L. E. Sweets
Department of Plant Pathology
Iowa State University
Ames, IA

A. T. Tschanz
U.S. Department of Agriculture
Animal and Plant Health Inspection
  Service
Hyattsville, MD

T. D. Wyllie
Department of Plant Pathology
University of Missouri
Columbia, MO

J. T. Yorinori
Plant Pathology
EMBRAPA/CNPSoja
Londrina, Paraná, Brazil

Contents

Introduction

The soybean, *Glycine max* (L.) Merrill, was domesticated by farmers in the eastern half of northern China during the Shang dynasty (ca. 1700–1100 B.C.) or perhaps earlier. For several thousand years people in eastern Asia have used soybeans for food and animal feed and as medicine to treat a number of human disorders.

The taxonomy of the genus *Glycine* is presented in Table 1. The geographic distribution suggests that the botanical home of *Glycine* may be Australia.

The soybean reached North America quite late. It was first introduced in 1765 by Samuel Bowen, a seaman employed by the East India Company. Bowen brought soybeans from China via London to Savannah (in the colony of Georgia). He manufactured soy sauce from soybeans grown on his farm and exported it to London.

In 1851, the soybean was introduced into the midwestern United States by Dr. Benjamin Franklin Edwards. He obtained seeds from Japanese fishermen who were rescued at sea by the ship *Auckland*, which was bringing sugar and other goods from Hong Kong to San Francisco. Edwards gave the seeds to John H. Lea, of Alton, Illinois, who planted them in his garden in the summer of 1851. By 1854, the soybeans brought to Illinois in 1851 had been multiplied, disseminated, and evaluated by farmers throughout the United States.

Today, soybeans are grown to some extent in most parts of the world and are a primary source of vegetable oil and protein. The oil is used primarily in margarine, salad oils, cooking oil, and shortening. In the United States, more than 90% of soybean meal, or the cake remaining after the oil has been extracted, is used to feed swine, poultry, fish, and household pets. Each year, soybean products become more important in the formulation of new, low-cost, nutritionally balanced, high-protein foods and beverages for human consumption.

Selected References

Hymowitz, T. 1987. Introduction of the soybean to Illinois. Econ. Bot. 41:28-32.

Hymowitz, T., and Harlan, J. R. 1983. The introduction of the soybean to North America by Samuel Bowen in 1765. Econ. Bot. 37:371-379.

Singh, R. J., Kollipara, K. P., and Hymowitz, T. 1988. Further data on the genomic relationships among wild perennial species (2n = 40) of the genus *Glycine* Willd. Genome 30:166-176.

(Prepared by T. Hymowitz)

Production

World oilseed production for 1987–1988 was estimated at a record 202 million tons, and soybeans accounted for half of the total. The United States produced 51.8 million tons, representing over 51% of world output, on about 58 million acres (23 million hectares). Approximately 37% of the U.S. crop is exported in the form of beans. Brazil had a record harvest of 18.5 million tons in 1987–1988. The People's Republic of China harvested 11.8 million tons during that period and is the third largest producer, followed by Argentina, with 8.5 million tons.

The bulk of the world's soybeans is processed into meal and oil. Soybean meal has benefited from the growing market preference for oilseeds with high meal and low oil contents. In 1987–1988 it accounted for about 60% of world output of oilseed meal. Most soybean meal produced in the United States, 61–62%, is used domestically; exports of it accounted for 69% of the total U.S. exports of oilseed meal in 1987–1988. About 26% of soybean exports from developed countries originated in the United States, and 20% in the European Community.

World production of vegetable oil reached 50 million tons in 1987–1988, of which soybean oil accounted for about 15 million tons, or 30% of the total. The developed countries are the major producers and exporters of soybean oil. In 1987–1988, these countries accounted for 60% of world production and supplied about 54% of world exports. The United States contributed 37% to world production and about one sixth of world exports. Brazil is the world's second largest producer of soybean oil and ranks third as exporter. The European Community is the world's leading soybean oil exporter, supplying nearly 37% of exports in 1987–1988.

(Prepared by S. C. Schmidt)

Table 1. Species in the Genus *Glycine* Willd.[a]

Species	Chromosome Number (2n)	Geographic Distribution
Subgenus *Glycine*		
G. *arenaria* Tind.	40	Australia
G. *argyrea* Tind.	40	Australia
G. *canescens* F. J. Herm.	40	Australia
G. *clandestina* Wendl.	40	Australia
G. *curvata* Tind.	40	Australia
G. *cyrtoloba* Tind.	40	Australia
G. *falcata* Benth.	40	Australia
G. *latifolia* (Benth.) Newell & Hymowitz	40	Australia
G. *latrobeana* (Meissn.) Benth.	40	Australia
G. *microphylla* (Benth.) Tind.	40	Australia
G. *tabacina* (Labill.) Benth.	40, 80	Australia, west central and South Pacific islands, Taiwan, (south China?)
G. *tomentella* Hayata	38, 40, 78, 80	Australia, Papua New Guinea, Philippines, Taiwan, (south China?)
Subgenus *Soja* (Moench) F. J. Herm.		
G. *soja* Sieb. & Zucc.	40	China, Taiwan, Japan, Korea, USSR
G. *max* (L.) Merrill	40	Cultigen (worldwide)

[a] Source: Singh et al, 1988.

Soybean Diseases

As soybean acreage has expanded throughout the world, soybean diseases have increased in number and severity. Some were previously undescribed; some are of unknown etiology; others were introduced from Asia or other production areas where they had been known for many years. In 1987, the world loss to soybean diseases was estimated at 10.3 million metric tons. Losses in the United States alone were estimated at $50–160 million in 1986 (estimates vary, depending upon market price).

Soybean diseases may be classified as infectious (biotic) or noninfectious (abiotic). Infectious diseases are caused by agents that can be transmitted from an infected to a healthy plant and cause disease when conditions are favorable for infection. Fungi, bacteria (including mycoplasmalike organisms), viruses, and nematodes cause infectious diseases. Various unfavorable environmental and nutritional conditions cause noninfectious diseases.

More than 100 pathogens are known to affect soybeans; about 35 are important economically. In addition, there are many reports of associations between soybeans and microorganisms in which the pathology is not understood. For example, protozoa in the genus *Phytomonas* were recently found in soybean seed tissues in Brazil. One or more diseases can generally be found in fields wherever soybeans are grown. All parts of the soybean plant are susceptible to a number of pathogens that reduce the quality and quantity of seed yields. The extent of losses depends upon the pathogen or condition involved, state of plant development and health when infection occurs, severity of the disease on individual plants, and number of plants infected or affected. In many diseases, a latent period elapses between infection and symptom expression. This latent period can range up to several weeks.

A pathogen may be very destructive one season and difficult or impossible to find the next season. Many pathogens can

Table 2. Growth Stage Key for Soybean Disease Evaluation[a,b]

Stage	Description[c]
V_1	Completely unrolled leaf at the unifoliolate node
V_2	Completely unrolled leaf at the first node above the unifoliolate node
V_3	Three nodes on the main stem, beginning with the unifoliolate node
V_N	N nodes on the main stem, beginning with the unifoliolate node
R_1	One flower at any node
R_2	Flower at node immediately below the uppermost node with a completely unrolled leaf
R_3	Pod 0.5 cm (0.25 in.) long at one of the four uppermost nodes with a completely unrolled leaf
R_4	Pod 2 cm (0.75 in.) long at one of the four uppermost nodes with a completely unrolled leaf
R_5	Beans beginning to develop (can be felt when the pod is squeezed) at one of the four uppermost nodes with a completely unrolled leaf
R_6	Pod contains full-sized green beans at one of the four uppermost nodes with a completely unrolled leaf
R_7	Pods yellowing, 50% of leaves yellow (physiologic maturity)
R_8	95% of pods brown (harvest maturity)

[a]Source: Fehr et al, 1971. Reproduced from Crop Science, Vol. 11, 1971, pp. 929-931, by permission of the Crop Science Society of America, Inc.

[b]To use the key to report disease occurrence at specific stages of soybean growth, first select a random sample of soybean plants for disease assessment. The sample may consist of individual leaves or plants, groups of plants, or all plants in a plot. Next, calculate the average infection for the units in the sample. Then, determine the vegetative stage by counting the number of nodes on the main stem (beginning with the unifoliolate node) that have a completely unrolled leaf and finding the corresponding entry in the table.

[c]See also Figure 2.

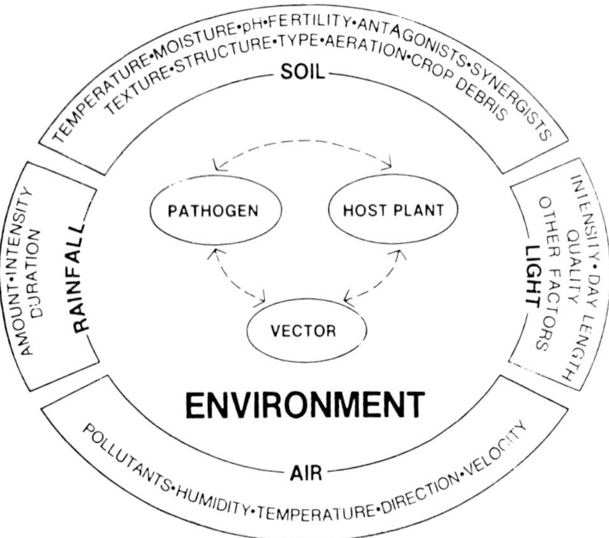

Fig. 1. Environmental influences on plant diseases caused by infectious agents.

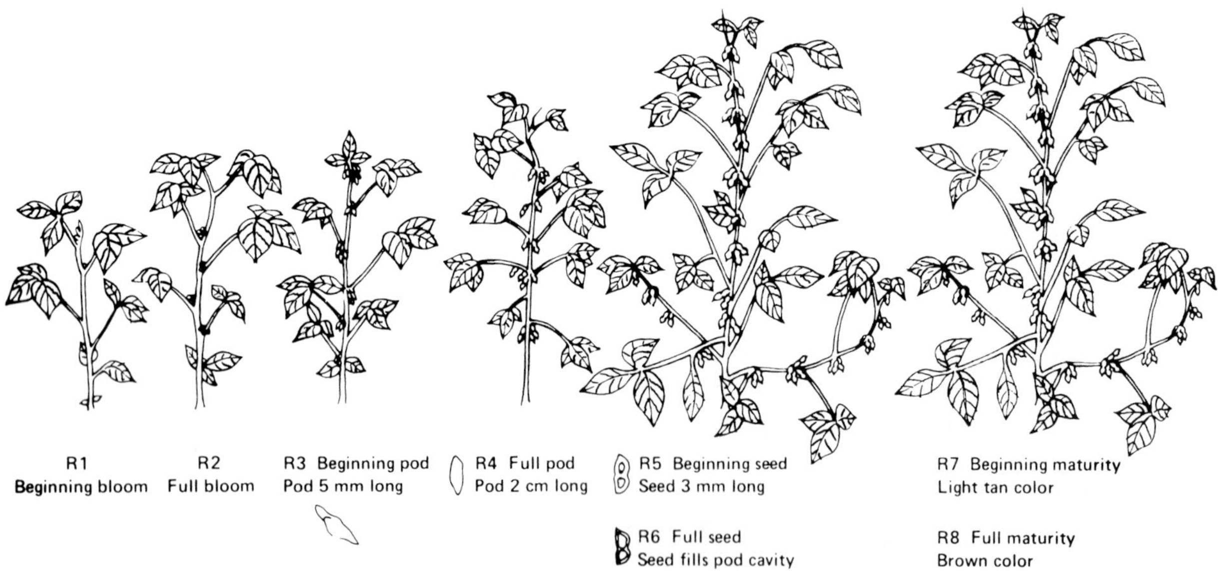

Fig. 2. Reproductive stages of soybeans. (Reprinted, by permission of the Southern Soybean Disease Workers, from W. J. Walla, ed., 1979, Soybean Diseases Atlas, 1st ed.)

or internationally via seeds.

When conditions are unfavorable for their growth and multi-plication, bacteria remain dormant on or in living or dead plants, soil, tools, equipment, or the bodies of insects or other animals. They generally survive best in the presence of other bacterial cells or host products. Most plant-pathogenic species die quickly at high temperatures (10 min at 51° C), under dry conditions, or under ultraviolet light (e.g., sunlight). Few pathogenic bacteria can survive in the soil; many are ingested by soil protozoa, and few have the biochemical functions necessary to compete for soil nutrients.

Selected References

Bradbury, J. F. 1986. Guide to Plant Pathogenic Bacteria. Common-wealth Agricultural Bureaux International, Farnham Royal, England.

Civerolo, E. L., Collmer, A., Davis, R. E., and Gillaspie, A. G., eds. 1987. Plant Pathogenic Bacteria. Martinus Nijhoff, Boston. 1,050 pp.

Dye, D. W., Bradbury, J. F., Goto, M., Hayward, A. C., Lelliott, R. A., and Schroth, M. N. 1980. International standards for naming pathovars of phytopathogenic bacteria and a list of pathovar names and pathotype strains. Rev. Plant Pathol. 59:153-168.

Krieg, N. R., ed. 1984. Bergey's Manual of Systematic Bacteriology, Vol. I, 9th ed. Williams & Wilkins, Baltimore.

Lelliott, R. A., and Stead, D. E. 1987. Methods for the Diagnosis of Bacterial Diseases of Plants. Blackwell Scientific Publications, Oxford. 216 pp.

Schaad, N. W., ed. 1988. Laboratory Guide for Identification of Plant Pathogenic Bacteria, 2nd ed. American Phytopathological Society, St. Paul, MN. 164 pp.

Sneath, P. H. A., Mair, N. S., and Sharpe, N. E., eds. 1986. Bergey's Manual of Systematic Bacteriology, Vol. II, 9th ed. Williams & Wilkins, Baltimore.

Starr, M. P., ed. 1982. Phytopathogenic Bacteria. Springer-Verlag, New York. 142 pp.

(Prepared by B. W. Kennedy and D. R. Hattermann)

Bacterial Blight

Bacterial blight occurs worldwide and is the most common bacterial disease of soybeans, especially during cool, wet weather. It occurs throughout the United States, where many cultivars are susceptible. Estimates of economic loss ranging from 5 to 18% were reported in Korea and the United States, but losses are generally very low where resistant or tolerant cultivars are used. The disease is most conspicuous in fields during mid-season. The causal bacterium remains active until checked by hot, dry weather.

Symptoms

Bacterial blight lesions are most conspicuous on leaves but also occur on stems, petioles, and pods. Small, angular, translucent, water-soaked, yellow to light brown spots appear on leaves (Plate 1). The centers soon dry out, turn reddish brown to black, and are surrounded by a water-soaked margin bordered by a yellowish green halo. Young leaves are most susceptible. Infected young leaves are distorted, stunted, and chlorotic. The angular lesions enlarge in cool, rainy weather and merge to produce large, irregular dead areas. The centers of older lesions frequently drop out or tear away, and thus leaves appear ragged, especially after strong winds and beating rains (Fig. 3). Early defoliation of lower leaves may occur. Large, black lesions develop on stems and petioles.

Commonly, lesions first appear on cotyledons, usually at the margins. The lesions enlarge and turn dark brown as the tissue collapses. Young seedlings may be stunted. If the shoot apical meristem is infected, the seedling usually dies.

Pod lesions, at first small and water-soaked, later enlarge and merge to involve much of the pod. They turn dark brown to black with age. Seeds become infected and may eventually be covered with a slimy bacterial growth. Stored seeds may shrivel, develop sunken or raised lesions, or become slightly discolored (Plate 2), or they may appear healthy. The bacterium elicits phytoalexin production in inoculated cotyledons.

Causal Organism

Pseudomonas syringae pv. *glycinea* (Coerper) Young, Dye, & Wilkie (syn. *P. glycinea* Coerper), the causal bacterium, is a motile, gram-negative rod, $1.2-1.5 \times 2.3-3.3$ μm, with rounded ends and one to several polar flagella. It weakly produces the green, water-soluble, fluorescent pigment characteristic of pseudomonads in culture. It does not liquefy gelatin and produces acid but no gas from sucrose. Colonies on nutrient agar are circular, smooth, and glistening, with an entire margin; they are white and raised but not viscid. The optimal temper-ature for growth is 24–26° C (the maximum is above 35° C; the minimum is above 2° C). The bacterium produces several toxins in culture and in vivo that are responsible for some of the symptomology. At least nine races of *P. syringae* pv. *glycinea* have been identified. Isolates resistant to antibiotics have been described in Brazil. A selective medium for isolation of the pathovar has been developed.

Disease Cycle and Epidemiology

P. syringae pv. *glycinea* overseasons in surface crop residue and in seeds. Seeds can be infected through the pods during the growing season, or they may be invaded during harvesting. Primary infections of cotyledons may be a major source of the inoculum that causes secondary lesions on seedlings. The bacterium is spread during windy rainstorms and during cultivation while the foliage is wet. It exists epiphytically on leaf surfaces and buds, needing only the proper temperature and windblown rain to enter the leaf.

The bacterium enters the plant through stomata and multiplies in the intercellular spaces of the mesophyll, where it produces a toxin that inhibits chlorophyll synthesis. Bacterial slime and fluids fill these spaces, and typical water-soaked lesions form within 5–7 days.

Fig. 3. Leaflet symptoms of bacterial blight of soybeans, caused by *Pseudomonas syringae* pv. *glycinea*. (Courtesy D. W. Chamberlain)

Strains of the bacterium infect bean (*Phaseolus vulgaris* L.), cowpea (*Vigna unguiculata* (L.) Walpers), lima bean (*P. lunatus* L.), and tepary bean (*P. acutifolius* A. Gray).

Cool, rainy weather favors the development of bacterial blight. Disease outbreaks usually follow windy rainstorms. The disease is seldom devastating to the crop.

Control

1. Avoid planting highly susceptible cultivars in areas where the disease is a potential problem.
2. Plant seeds relatively free of the pathogen.
3. Rotate soybeans with crops not susceptible to the pathogen.
4. Completely cover plant residue by clean plowing after harvest.
5. Do not cultivate when the foliage is wet.
6. Antibiotic sprays have not been successful in controlling bacterial blight, and copper-based materials have been inadequately tested.

Selected References

Fieldhouse, D. F., Sasser, M., Burbage, D. A., and Mucha, C. F. 1984. An improved selective medium for isolation of *Pseudomonas syringae* pv. *glycinea* from seed, soil and plant residue. (Abstr.) Phytopathology 74:756.

Holliday, M. J., and Keen, N. T. 1982. The role of phytoalexins in the resistance of soybean leaves to bacteria: Effect of glyphosate on glyceollin accumulation. Phytopathology 72:1470-1474.

Kennedy, B. W. 1980. Transmission of *Pseudomonas glycinea* via soybean seeds. Page 79 in: Abstr. World Soybean Res. Conf. II. F. T. Corbin, ed. Westview Press, Boulder, CO.

Kennedy, B. W., and Ercolani, G. L. 1978. Soybean primary leaves as a site for epiphytic multiplication of *Pseudomonas glycinea*. Phytopathology 68:1196-1201.

Mew, T. W., and Kennedy, B. W. 1982. Seasonal variation in populations of pathogenic pseudomonads on soybean leaves. Phytopathology 72:103-105.

Park, E. W., and Lim, S. M. 1985. Overwintering of *Pseudomonas syringae* pv. *glycinea* in the field. Phytopathology 75:520-524.

Park, E. W., and Lim, S. M. 1986. Effects of bacterial blight on soybean yield. Plant Dis. 70:214-217.

(Prepared by B. W. Kennedy and J. B. Sinclair)

Bacterial Pustule

Bacterial pustule has been reported in most soybean-growing areas of the world where warm weather and frequent showers prevail during the growing season. It causes premature defoliation, which may decrease yield by reducing seed size and number.

Symptoms

Early symptoms are minute, pale green spots with elevated centers on either or both leaf surfaces. Later, a small, raised, light-colored pustule forms in the center, usually in lesions on the underleaf surface (Fig. 4 and Plates 1 and 3). The pustules form through hypertrophy and hyperplasia. These symptoms are sometimes confused with those of soybean rust (see Rust).

The spots vary from specks to large, irregular, mottled brown areas, which develop when lesions coalesce. Leaves become ragged when dead areas are torn away by wind. Severe infection often results in some defoliation. Leaf spots sometimes form without developing pustules. Small, reddish brown, slightly raised spots may develop on pods of susceptible cultivars.

Symptoms of bacterial pustule sometimes resemble those of bacterial blight. However, pustule lesions are not water-soaked in the early stages of development and usually have minute, raised pustules in the centers, especially if viewed from the lower surface.

Phytoalexins are produced in response to infection.

Causal Organism

Xanthomonas campestris pv. *glycines* (Nakano) Dye (syns. *X. campestris* pv. *phaseoli* (Smith) Dye and *X. phaseoli* (Smith) Dowson var. *sojensis* (Hedges) Starr & Burkholder), the causal agent of bacterial pustule, is a motile, gram-negative rod, 0.5–0.9 × 1.4–2.3 μm, with a single polar flagellum. Colonies on beef infusion agar are pale yellow (becoming deeper yellow with age), small, circular, and smooth, with an entire margin. The bacterium liquefies gelatin, produces acid but no gas from sucrose, and rapidly hydrolyzes starch. The optimal temperature for growth is 30–33°C (the maximum is 38°C; the minimum is 10°C). The bacterium produces auxins, bacteriocin, and exopolysaccharides in culture.

Disease Cycle and Epidemiology

The bacterium overseasons in soybean seeds (see Other Seedborne Bacteria, under Seed Pathology), in surface crop debris, and in the rhizosphere of wheat roots. Weeds, such as redvine (*Brunnichia cirrhosa* Gaertn.) in the United States and *Dolichos biflorus* L. in India, also are hosts. Strains infect beans (*Phaseolus*) and cowpea (*Vigna*). The bacterium spreads via splashing water or windblown rain and during cultivation when foliage is wet.

The bacterium enters the plant through natural openings and wounds and multiplies intercellularly. Bacterial pustule, unlike bacterial blight, is not checked by high temperatures. New infections may occur throughout the growing season whenever wet or rainy conditions prevail. The disease can be severe in the tropics.

The *rxp* gene confers resistance in soybeans by increasing the number of bacterial cells necessary for infection.

Control

1. Use resistant cultivars.
2. Follow the control measures suggested for bacterial blight.

Selected References

Groth, D. E., and Braun, E. J. 1986. Growth kinetics and histopathology of *Xanthomonas campestris* pv. *glycines* in leaves of resistant and susceptible soybeans. Phytopathology 76:959-965.

Jones, S. B., and Fett, W. F. 1987. Bacterial pustule disease of soybean: Microscopy of pustule development in a susceptible cultivar. Phytopathology 77:266-274.

Laviolette, F. A., Athow, K. L., Probst, A. H., and Wilcox, J. R. 1976. Effect of bacterial pustule on yield of soybeans. Crop Sci. 10:150-151.

(Prepared by B. W. Kennedy and J. B. Sinclair)

Fig. 4. Bacterial pustule lesions, caused by *Xanthomonas campestris* pv. *glycines*, on the lower surface (left) and upper surface (right) of soybean leaflets. (Courtesy U.S. Department of Agriculture)

Bacterial Tan Spot

Bacterial tan spot has been reported in Canada, the Soviet Union, and the United States. It was first observed in the United States in 1975. In a three-year study in Iowa, the disease reduced yields by 13% one year but caused no reduction the next. Bruising of leaves by hail resulted in rapid disease spread, which reduced seed yield by 17%.

Symptoms

Infection begins with a small lesion, which enlarges until the entire leaflet is involved (Fig. 5). Chlorosis begins in oval or elongated patterns, frequently along leaf margins, and progresses inward toward the midrib. Infected tissues die, dry out, and turn brown. During moist periods, fungi may invade the dead tissues, discoloring them. Tan areas may fall out under windy conditions, so that leaves acquire a ragged appearance. Because fruiting structures of saprophytic fungi sometimes form in tan spot lesions, and because the lesions are similar to those of Phyllosticta leaf spot, the two diseases can only be identified by isolation of the pathogen.

When infected seeds are sown, the tan spot bacterium may cause seedlings to be stunted and have fused leaflets and, later, empty pods. Leaves may develop marginal necrosis or necrotic spots on interior portions of the lamina. Infection is reduced after flower set.

Causal Organism

Curtobacterium flaccumfaciens pv. *flaccumfaciens* (Hedges) Collins & Jones causes both wilt of bean (*Phaseolus vulgaris* L.) and tan spot of soybeans. The wilt strain from bean does not produce tan spot symptoms on soybeans, but the tan spot strain from soybeans produces bean wilt. A useful character for identification is the acetyl cell wall structure with a D-ornithine peptidoglycan. Young cells are small, irregular rods and usually react gram-positively; however, older or younger cells may give a gram-negative reaction. Cells become shorter to coccoid in older cultures and generally are motile with lateral flagellation. Colonies on beef agar are yellow, circular, smooth, slightly convex, opaque, and slightly viscid, with an entire margin. They appear wet and shiny. The optimum growth temperature is 24–27°C, and the maximum temperature 35–37°C.

Disease Cycle and Epidemiology

C. flaccumfaciens pv. *flaccumfaciens* is internally seedborne in beans and soybeans. When infected soybean seeds are sown, infection of unifoliolate and the first trifoliolate leaves may occur via the vascular system at 25–30°C (Fig. 5). The spread of tan spot from infected seedling leaves is slow (2.2 m per week from a single infected plant), because it is dependent upon wounds resulting from leaf-to-leaf contact caused by wind. Crop damage is a function of the number of seedlings initially infected in a field, which is determined by the percentage of diseased seeds sown. Cultivars in which up to 30% of seeds are infected have been recorded.

Control

1. Use resistant cultivars.
2. Plant uninfected seeds.
3. If seeds are infected, plant early, when temperatures tend to be below 25°C, to escape leaf infection via the vascular system.

Selected References

Dunleavy, J. M. 1984. Yield losses in soybeans caused by bacterial tan spot. Plant Dis. 68:774-776.

Dunleavy, J. M. 1985. Spread of bacterial tan spot of soybean in the field. Plant Dis. 69:1036-1039.

Dunleavy, J. M., Keck, J. W., Gobelman, K. S., Reddy, S., and Thompson, M. M. 1983. Prevalence of *Corynebacterium flaccumfaciens* as incitant of bacterial tan spot of soybean in Iowa. Plant Dis. 67:1277-1279.

(Prepared by J. M. Dunleavy)

Wildfire

Wildfire, a damaging disease of tobacco (*Nicotiana* spp.), has been reported in soybeans in the United States and Brazil. Its present significance is minor.

Symptoms

The wildfire bacterium causes brown, necrotic spots on leaves. The spots vary in size and shape and are nearly always surrounded by a broad, yellow halo (Fig. 6 and Plate 4).

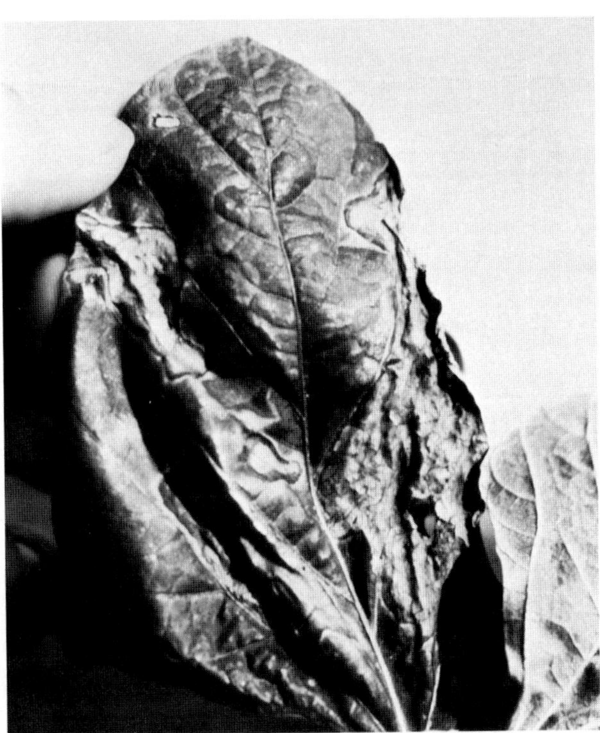

Fig. 5. Bacterial tan spot, caused by *Curtobacterium flaccumfaciens* pv. *flaccumfaciens*. (Courtesy J. M. Dunleavy)

Fig. 6. Wildfire lesions, caused by *Pseudomonas syringae* pv. *tabaci*, on soybean leaflets. (Courtesy U.S. Department of Agriculture)

Restricted spots may occur without the sharply delineated chlorotic zones. These spots are dark brown to black, in contrast to the usual light brown, more expansive type of lesion. In damp weather, lesions enlarge and coalesce to form large dead areas, which become dry and tear away, so that leaves become tattered. When the disease is severe, the plant may defoliate prematurely.

Bacterial pustule lesions are natural infection courts for the wildfire bacterium. A pustule formed by *Xanthomonas campestris* pv. *glycines*, the bacterial pustule pathogen, can usually be found in the center of a wildfire lesion.

Causal Organism

Pseudomonas syringae pv. *tabaci* (Wolf & Foster) Young, Dye, & Wilkie (syn. *P. tabaci* (Wolf & Foster) F. L. Stevens), the causal agent of wildfire, is a motile, gram-negative rod, $0.5–1.2 \times 1.4–3.3$ μm, with a polar flagellum. It produces a green, water-soluble, fluorescent pigment in culture, liquefies gelatin, and produces acid but not gas from sucrose. Colonies on nutrient agar are white, circular, and slightly raised, with a translucent margin. The optimal temperature for growth is 24–28°C (the maximum is 38°C; the minimum is about 4°C).

Isolates of the bacterium from soybeans are morphologically, physiologically, serologically, and pathologically similar to isolates from tobacco. Tobacco isolates are generally more virulent on tobacco than on soybeans, but soybean isolates are equally virulent on both. Except for pathogenicity, *P. syringae* pv. *tabaci* resembles *P. fluorescens* Migula, a common soil saprophyte.

Disease Cycle and Epidemiology

The bacterium may overseason in infected plant debris and seeds. It grows on root surfaces of many crop plants and is spread by splashing water and windblown rain.

Water congestion of plant tissues caused by beating rains is often required for invasion and infection. Seasonal temperature variations kill the bacterium and may influence its geographic distribution.

Control

1. Use cultivars resistant to both the bacterial pustule and the wildfire bacteria.
2. Follow the control measures suggested for bacterial blight.

Selected Reference

Ribeiro, R. de L. D., Hagedorn, D. J., Durbin, R. D., and Uchytil, T. F. 1979. Characterization of the bacterium inciting bean wildfire in Brazil. Phytopathology 69:208-212.

Bacterial Wilts

Two bacterial wilts have been described in the United States and the Soviet Union. Neither disease is common in nature. Most reports from the United States refer to artificial inoculation with bacteria known to be wilt-inducing; those from the Soviet Union are from field infections. Two bacteria were reported to produce wilting in soybeans: *Curtobacterium flaccumfaciens* pv. *flaccumfaciens* (Hedges) Collins & Jones, in Iowa and in the Priamur region of the Soviet Union, and *Pseudomonas solanacearum* (Smith) Smith, in North Carolina and in the Ukrainian Republic. Both are important pathogens of other host species. Pseudomonas wilt occurs mainly in the western regions of the Soviet Union and can cause considerable damage.

Symptoms

C. flaccumfaciens pv. *flaccumfaciens* causes a progressive wilt with leaf yellowing and vascular discoloration. Leaves die,

dry out, and remain on the plant. Infected plants are weak and stunted and eventually die.

The leaves of plants infected with *P. solanacearum* become chlorotic and then develop small, dark brown necrotic spots, which elongate and develop dark borders. The spreading leaf lesions dry out, and affected tissues drop out, so that leaves acquire a ragged appearance. A progressive withering follows, involving the entire plant. Some young plants may develop a rapid, severe wilt; others wilt only slightly and become stunted.

Causal Organisms

C. flaccumfaciens pv. *flaccumfaciens* (syns. *Corynebacterium flaccumfaciens* (Hedges) Dowson and *Corynebacterium flaccumfaciens* pv. *flaccumfaciens* (Hedges) Dowson) is a motile, gram-positive rod, $0.3–0.5 \times 0.6–3.0$ μm, with a polar flagellum. Colonies on beef agar are yellow, circular, smooth, flat or slightly convex, opaque, and slightly viscid, with an entire margin. They appear wet and shiny. The optimal temperature for growth is 31°C (the maximum is 36–40°C; the minimum is below 1.5°C).

P. solanacearum is a motile, gram-negative rod, 0.5×1.5 μm, with a polar flagellum. Colonies on agar are small, opalescent (becoming brown), irregular, and smooth; they appear wet and shiny. The optimal temperature for growth is 35–37°C (the maximum is about 41°C; the minimum is about 10°C).

Disease Cycle and Epidemiology

C. flaccumfaciens pv. *flaccumfaciens* is seedborne; *P. solanacearum* is occasionally seedborne. Both may overseason in crop residue. Common hosts of *P. solanacearum* include banana, pea, peanut, potato, tobacco, and tomato. Garden bean is the principal host of *C. flaccumfaciens* pv. *flaccumfaciens*. The disease caused by the latter is favored by low temperatures and abundant rainfall, and soil conditions are usually poor where the disease occurs.

Control

1. Use sanitation practices.
2. Follow the control measures suggested for bacterial blight.

Selected References

Dunleavy, J. B. 1963. A vascular disease of soybeans caused by *Corynebacterium* sp. Plant Dis. Rep. 47:612-613.
Hedges, F. 1926. Bacterial wilt of beans (*Bacterium flaccumfaciens* Hedges), including comparisons with *Bacterium phaseoli*. Phytopathology 16:1-22.
Miakushko, Iu. P., and Baranova, V. F. 1984. Soia. Kolos, Moscow. 332 pp.
Podkina, D. V., Nikitina, K. V., Belekhova, K. A., and Andreeva, L. T. 1980. Bacterial diseases of soybeans in the Krasnodar region. (In Russian) Byull. Inst. Rastenievod. imeni N. I. Vavilova 97:70-73.

(Prepared by B. A. Fulling and J. B. Sinclair)

Crown Gall

Crown gall bacteria are used for biotechnological research on many plants, including soybeans. Induction of crown gall tumors or hairy root disease has been achieved under artificial conditions by inoculation with *Agrobacterium rhizogenes* (Riker et al) Conn or *A. tumefaciens* (Smith & Townsend) Conn. Tumors have also been induced by *A. tumefaciens* on *Glycine soja* Sieb. & Zucc., in addition to many cultivars of *G. max*.

Selected References

Owens, L. 1984. Genotypic variability of soybean response to *Agrobacterium tumefaciens* and *A. rhizogenes*. (Abstr.) Plant Physiol.

(Suppl.) 75:31.

Pedersen, H., Christiansen, J., and Wyndaele, R. 1983. Induction and
in vitro culture of soybean crown gall tumours. Plant Cell Rep.
2:201-204.

Wang, L. Z., Yin, G. C., Luo, J. F., Lei, B. J., Wang, J., Yao, Z. C., Li,
X. L., Shae, Q. Q., Jiang, X. C., and Zhou, Z. Q. 1984. A study on
tumor formation of soybean and gene transfer. Sci. Sin. B 27:391-397.

Other Bacteria

Soybeans have been reported to be susceptible to isolates of
the following phytobacteria: *Erwinia amylovora* (Burr.)
Winslow, Broadhurst, Buchanan, Krumwiede, Rogers, & Smith;
E. carotovora var. *atroseptica* (van Hall) Dye; *Pseudomonas
andropogonis* (Smith) Stapp; *P. syringae* pv. *phaseolicola*
(Burkholder) Young, Dye, & Wilkie; *P. syringae* pv. *syringae*
van Hall; and *Xanthomonas campestris* pv. *cannabis* Severin.

Selected References

Garcia, J. R., and Zak, L. F. 1983. Effect of various bacteria on the
germination and seedling emergence of soybean. (In Spanish)
Agrocienc. Mex. 51:69-80.
Nikitina, K. V., and Korsakov, N. I. 1978. Bacterial diseases of soybean
in the Soviet Far East and in southern regions of the U.S.S.R.: Search
for sources of resistance to them. (In Russian) Tr. Prikl. Bot. Genet.
Sel. 62:13-18.
Severin, V. 1978. A new pathogenic bacterium on hemp—*Xanthomonas
campestris* pv. *cannabis*. (In German) Arch. Phytopathol. Pflanzen-
schutz 14:7-15.
Stall, R. E., and Kucharek, T. A. 1982. A new bacterial disease of
soybean in Florida. (Abstr.) Phytopathology 72:990.

Mycoplasmalike Diseases

Mycoplasmalike organisms (MLOs), or mycoplasmas, are
bacteria that lack a firm outer wall. Another group, called
spiroplasmas, lack a cell wall, are bounded by a trilaminar
membrane, and grow as mobile, helical, filamentous cells.

MLOs commonly occur in the food-conducting tissue of
plants. Most diseases thought to be caused by MLOs or
Spiroplasma spp. are the so-called yellows types. Characteristic
symptoms of infection by MLOs include yellowing, shortening
of internodes, phyllody, proliferation of axillary shoots to
produce witches'-broom, sterility, virescence, and reduction in
root growth. MLOs are suspected of causing several soybean
diseases, including machismo, bud proliferation, phyllody, and
witches'-broom.

ever, tetracycline causes only a temporary remission of symptoms. Control measures include controlling weed hosts and
roguing infected plants.

Selected References

Fletcher, J., Irwin, M. E., Bradfute, O. E., and Granada, G. A. 1984. A
machismo-like disease of soybeans in Mexico. Phytopathology
74:857.
Granada, G. A. 1979. Machismo disease of soybeans: I. Symptom-
atology and transmission. Plant Dis. Rep. 63:47-50.
Granada, G. A. 1979. Machismo disease of soybeans: II. Suppressive
effects of tetracycline on symptom development. Plant Dis. Rep.
63:309-312.

Machismo

Machismo was reported in Colombia in 1968 and Mexico in
1984. It has been economically important in Colombia since
1974, with reported yield losses of up to 80%. Besides soybeans,
field bean (*Phaseolus vulgaris* L.), species of *Crotalaria* and
Desmodium, *Galactia glaucescens* H.B.K., pigeon pea (*Cajanus
cajan* (L.) Millsp.), *Rhynchosia minima* DC., and *Vinca rosea*
L. are infected.

Symptoms of machismo vary with the host, cultivar, and age
of plants when infected. The first symptoms usually appear
about the time of flowering or after pod formation begins. Pods
may be rigid, curved, flat, and thin, usually in an upright
position, with no beans formed. Alternatively, they may be
transformed into corrugated leaflike structures (Fig. 7), in a
condition resembling phyllody. Once the floral organs have
been transformed into leaflike structures, buds proliferate from
leaf axils anywhere on the plant, causing witches'-broom. The
younger the plant when infected, the more severe the symptoms.

Sepals of infected plants may be longer, larger, and stronger
than healthy sepals, giving the appearance of small leaves. The
color intensity of petals may be diminished. Flowers are smaller
and remain closed; in these cases, the podlike or leaflike
structures grow through the tips of the unopened flowers.
Often, part of a plant shows symptoms while the remainder
appears healthy, with normal development of flowers, pods,
and beans. When this occurs, the affected branch is usually
close to the bottom of the plant.

The causal agent is transmitted by the leafhopper *Scaphytopius fuliginosus* Osborn. The disease has not been transmitted by mechanical inoculation, nor is there evidence of seed
transmission.

Tetracycline treatments of plants in the early stages of the
disease have prevented further symptom development. How-

Fig. 7. Machismo, caused by a mycoplasmalike organism.
(Courtesy M. E. Irwin)

Bud Proliferation

Bud proliferation, which resembles machismo, has been reported in Louisiana. The disease is seed-transmitted and spread by the leafhopper *Scaphytopius acutus* (Say). It is not mechanically transmitted.

Selected Reference

Derrick, K. S., and Newsom, L. D. 1984. Occurrence of a leafhopper-transmitted disease of soybeans in Louisiana. Plant Dis. 68:343-344.

Witches'-Broom and Phyllody

Witches'-broom and phyllody resemble machismo. They cause a phylloid disorder of the floral organs, shoot proliferation, and reduced leaflets. Plants with these symptoms have been reported in Brazil, India, Indonesia, Japan, Mozambique, Nigeria, and Tanzania. No comprehensive report is available.

The leafhopper vectors of witches'-broom are an *Orosius* sp. in Brazil, *O. orientalis* (Matsumura) in Japan, and *O. argentatus* (Evans) in India and Indonesia.

In India, symptoms of witches'-broom were suppressed when soybean plants were treated with tetracycline.

Selected References

Ebbels, D. L., and Allen, D. J. 1979. A Supplementary and Annotated List of Plant Diseases, Pathogens and Associated Fungi in Tanzania. Phytopathol. Pap. 22. Commonwealth Mycological Institute, Kew, Surrey, England. 89 pp.

Tarp, G., Lange, L., and Kongsdal, O. 1987. Seed-borne pathogens of major food crops in Mozambique. Seed Sci. Technol. 15:793-810.

Beneficial Bacteria

Soybean roots can be colonized by nitrogen-fixing bacteria in the genera *Bradyrhizobium* and *Rhizobium*. They form a symbionic relationship that is beneficial to the plant by fixing or trapping atmospheric nitrogen and converting it into an organic form. The bacteria colonize roots and induce the formation of well-organized structures, called nodules, on the roots. Some strains of the bacteria parasitize roots but are not capable of fixing nitrogen. Therefore, it cannot be assumed that the number of nodules present is indicative of nitrogen-fixing activity. When inorganic nitrogen is supplied at levels that meet plant needs, the activity of the bacteria tends to be suppressed.

Nodules

Nodules develop after direct penetration of root hairs or epidermal cells by the bacteria. The nodules vary in size and shape, depending upon many factors, including the strain of the bacterium, time and place of infection, and soil conditions. Generally, nodules range from 1 mm to 2–3 cm in diameter; they may be cylindrical, round, or irregularly shaped and rough or smooth. A mature nodule consists of an epidermal layer, a cortical layer, and a central region, which contains bacteroids, or membrane-enclosed bacteria. The cortical tissues contain vascular elements, which are not in contact with those of the host root. A meristematic region develops around the nodule or at the tip. The color of nodules varies from whitish to cream to brown, depending upon age and soil conditions.

Causal Organisms

The bacteria are gram-negative, aerobic, polar or subpolar flagellate, and generally rod-shaped (1.2–3.0×0.5–0.9 μm), although they can vary from irregular to club-shaped. *B. japonicum* (Kirchner) Jordan, a slow-growing type, is the most common species on soybeans. *R. japonicum* (Kirchner) Buchanan (which includes *R. loti*), a fast-growing type, is found only on some soybean cultivars from the People's Republic of China.

Both species nodulate on roots of species of *Cicer*, *Lablab*, *Leucaena*, *Mimosa*, and *Sesbania*. Isolates or strains within either species show crop preference and even cultivar preference within a crop. For example, some strains work better on some soybean cultivars than on others.

The bacteria survive in tissues of susceptible hosts and in the soil. Their survival depends upon soil conditions and whether the particular host or cultivar is planted continuously in the field.

Selected Reference

Jordan, D. C. 1982. Transfer of *Rhizobium japonicum* to *Bradyrhizobium*, a slow-growing root nodule bacterium from leguminous plants. Int. J. Syst. Bacteriol. 32:136-139.

Fungal Diseases

Most plant diseases are caused by fungi. Of the nearly 70,000 described species of fungi, more than 8,000 are known plant pathogens. Numerous additional fungus-caused diseases of plants are discovered every year.

The fungi are a diverse group of plantlike organisms that lack chlorophyll and cannot carry out photosynthesis. Instead, they obtain sustenance from organic matter produced by other organisms. Fungi are adapted for survival in air, soil, and water and typically are in close association with other organisms. Because they do not require sunlight, fungi can live in darkness as well as in the light.

Fungi acquire nutrients in a number of different ways. Saprophytic fungi use nonliving organic matter (such as dead leaves or stems, animal carcasses, and construction materials derived from natural substances) for food. Parasitic fungi require a living host for obtaining nutrients. A large percentage of fungi can function either as saprophytes or as parasites. In general, obligately parasitic fungi tend to have narrower host ranges and to be more closely adapted to the biology of their hosts than fungi capable of growing saprophytically.

Most fungi produce hyphae, which are microscopic, cellular, threadlike filaments. Hyphae grow and branch to form mycelium, which is the vegetative body of the fungus. Mycelial structures range in size from microscopic wefts of hyphae to large mushrooms and conks.

Typically, fungi reproduce and spread by means of spores. Some spores, as well as larger structures such as sclerotia, are adapted for survival in the absence of a host. Asexual spores are generally produced in great numbers during the growing season and are often responsible for the development of epidemics. Sexual spores are the result of genetic recombination and therefore serve to maintain genetic diversity in fungal populations. Spores can be disseminated by air currents, splashing or flowing water, and the activities of people and animals. In some diseases, insects or other animals are important not only as agents of dissemination but as agents of inoculation.

Fungal taxonomy is based largely on the morphology of spores, the structures that produce them, and their development. Other characteristics, such as host specificity, also are used to classify fungi. Fungal taxonomy is changing rapidly, and in certain fungal groups, information on hyphal anastomosis groups, isozyme variation, restriction-fragment-linked polymorphisms, and other new kinds of taxonomic characters are being used to improve the system by which fungi are identified and classified. In the future, such advances will likely improve methods for diagnosing the causes of plant diseases. A classification of the fungi reported occur in

Table 4. Classification of Fungi Reported to Occur in Soybeans[a,b]

Division: Eumycota	Subdivision: Ascomycotina	Subdivision: Deuteromycotina
Subdivision: Mastigomycotina	(*continued*)	(*continued*)
Class: Chytridiomycetes	Class: Pyrenomycetes (*continued*)	Class: Hyphomycetes
Order: Chytridiales	Order: Sordariales	*Acremonium* sp.
Olpidium viciae	*Ophioceras* sp.	**Alternaria* sp.
Synchytrium dolichi	Order: Polystigmatales	**Alternaria alternata*
Class: Oomycetes	**Glomerella cingulata*	**Alternaria atrans*
Order: Peronosporales	**Glomerella glycines*	**Alternaria tenuissima*
**Peronospora manshurica*	Order: Hypocreales	*Bipolaris victoriae*
**Phytophthora megasperma*	**Calonectria crotalariae*	*Botrytis cinerea*
f. sp. *glycinea*	**Gibberella zeae*	*Cercospora canescens*
**Pythium aphanidermatum*	**Neocosmospora vasinfecta*	**Cercospora kikuchii*
**Pythium debaryanum*	*Neocosmospora vasinfecta*	**Cercospora sojina*
**Pythium irregulare*	var. *africana*	*Chaetoseptoria wellmanii*
**Pythium myriotylum*	Subdivision: Basidiomycotina	*Clasterosporium* sp.
**Pythium ultimum*	Class: Urediniomycetes	**Corynespora cassiicola*
Subdivision: Zygomycotina	Order: Uredinales	**Cylindrocladium clavatum*
Class: Zygomycetes	**Phakopsora pachyrhizi*	**Cylindrocladium crotalariae*
Order: Mucorales	*Uromyces phaseoli* var. *typica*	**Cylindrocladium floridanum*
**Choanephora cucurbitarum*	Class: Hymenomycetes	**Cylindrocladium scoparium*
**Choanephora infundibulifera*	Order: Aphyllophorales	*Dendryphion* sp.
Mucor sp.	**Thanatephorus cucumeris*	**Drechslera glycini*
Subdivision: Ascomycotina	Order: Auriculariales	*Drechslera halodes*
Class: Discomycetes	*Helicobasidium mompa*	*Fusarium acuminatum*
Order: Helotiales	*Stilbum* sp.	*Fusarium bulbigenum*
**Sclerotinia sclerotiorum*	Subdivision: Deuteromycotina	var. *tracheiphilum*
Class: Loculoascomycetes	Class: Agonomycetes	**Fusarium equiseti*
Order: Dothideales	*Rhizoctonia centrifugum*	**Fusarium graminearum*
Leptosphaerulina briosiana	*Rhizoctonia sasakii*	**Fusarium moniliforme*
Leptosphaerulina trifolii	**Rhizoctonia solani*	**Fusarium moniliforme*
Mycosphaerella sp.	**Sclerotium rolfsii*	var. *subglutinans*
Mycosphaerella cruenta	Class: Coelomycetes	**Fusarium orthoceras*
Mycosphaerella phaseolorum	*Aristastoma camarographioides*	**Fusarium oxysporum*
Mycosphaerella sojae	*Ascochyta* sp.	**Fusarium oxysporum*
**Mycosphaerella uspenskajae*	*Ascochyta phaseolorum*	f. sp. *tracheiphilum*
Pleosphaerulina (= *Pringsheimia,*	**Ascochyta sojae*	**Fusarium oxysporum*
fide von Arx & Mueller) *glycines*	*Ascochyta sojicola*	f. sp. *vasinfectum*
Pleosphaerulina (= *Pringsheimia,*	*Botryodiplodia pallida*	**Fusarium rigidiusculum*
fide von Arx & Mueller) *sojicola*	*Colletotrichum circinans*	*Fusarium roseum*
Order: Chaetothyriales	**Colletotrichum destructivum*	**Fusarium semitectum*
Trotteria (= *Actinopeltis,*	**Colletotrichum gloeosporioides*	**Fusarium solani*
fide Sutton) *venturioides*	**Colletotrichum graminicola*	**Fusarium tricinctum*
Order: Pleosporales	**Colletotrichum truncatum*	*Fusarium udum* f. sp. *crotalariae*
Guignardia sojae	*Coniothyrium sojae*	*Gliocladium roseum*
Pleospora herbarum	**Dactuliochaeta glycines*	*Isariopsis griseola*
Class: Plectomycetes	**Dactuliophora glycines*	*Mycoleptodiscus terrestris*
Order: Eurotiales	*Macrophoma mame*	*Myrothecium* sp.
Eurotium sp.	**Macrophomina phaseolina*	*Myrothecium roridum*
Class: Pyrenomycetes	*Marssonina sojicola*	**Phialophora glycines*
Order: Erysiphales	*Phoma exigua*	*Phymatotrichopsis omnivorum*
**Microsphaera diffusa*	**Phomopsis longicolla*	*Pseudocercospora cruenta*
Order: Diaporthales	*Phomopsis phaseoli*	*Stemphylium botryosum*
**Diaporthe phaseolorum*	**Phomopsis sojae*	*Stemphylium vesicarium*
var. *caulivora*	*Phyllosticta glycinea*	**Thielaviopsis basicola*
**Diaporthe phaseolorum* var. *sojae*	**Phyllosticta sojicola*	*Trichoderma* sp.
Gaeumannomyces graminis	*Pseudorobillarda sojae*	*Verticillium albo-atrum*
var. *graminis*	*Septogloeum sojae*	*Verticillium dahliae*
Order: Sphaeriales	**Septoria glycines*	
Rosellinia sp.	**Sphaceloma glycines*	

[a] Based primarily on Hawksworth et al, 1983, and Barr, 1987.
[b] * = Known to be pathogenic.

soybeans naturally or by inoculation is presented in Table 4.

Many fungi enter soybean plants through natural openings, such as stomata, hydathodes, nectaries, and lenticels. In other cases they infect plants through wounds made by blowing sand, wind, hail, people, insects, equipment, nematodes, or other agents. Hyphae or specialized structures such as penetration pegs generally penetrate plant hosts through a combination of enzyme action and pressure. Fungi can overseason on or in living or dead plants, seeds, soil, and occasionally insects. Some fungi, in addition to growing in soybean plants, infect and reproduce in weeds or other hosts, where they can also overseason.

Physiologic races or strains of fungi are morphologically indistinguishable but differ in characteristics such as the species, varieties, and cultivars of host plants that they can parasitize and the degree of virulence they exhibit.

Many fungi colonize soybean plants and seed tissues asymptomatically. Latently infected plants often develop symptoms if they are under stress or near the end of the growing season, when they are senescing. In recent years, scientists have recognized that many fungi are present as endophytes—that is, they are present in symptomless plants and do not cause disease. Members of the following genera cause latent infections in soybeans: *Cercospora, Colletotrichum, Diaporthe* (*Phomopsis*), *Fusarium*, and *Macrophomina*.

Selected References

Agrios, G. N. 1988. Plant Pathology, 3rd ed. Academic Press, New York. 803 pp.
Alexopoulos, C. J., and Mims, C. W. 1979. Introductory Mycology, 3rd ed. John Wiley & Sons, New York. 632 pp.
Barr, M. E. 1987. Prodromus to Class Loculoascomycetes. Hamilton I. Newell, Amherst, MA. 168 pp.
Cerkauskas, R. F., and Sinclair, J. B. 1980. Use of paraquat to aid detection of fungi in soybean tissues. Phytopathology 70:1036-1038.
Hawkworth, F. L., Sutton, B. C., and Ainsworth, G. C. 1983. Dictionary of the Fungi, 7th ed. Commonwealth Mycological Institute, Kew, Surrey, England. 445 pp.
Rossman, A. Y., Palm, M. E., and Spielman, L. J. 1987. A Literature Guide for the Identification of Plant Pathogenic Fungi. American Phytopathological Society, St. Paul, MN. 252 pp.
Webster, J. 1980. Introduction to Fungi, 2nd ed. Cambridge University Press, Cambridge. 669 pp.

(Prepared by D. A. Glawe)

Fungal Diseases of Foliage, Upper Stems, Pods, and Seeds

Alternaria Leaf Spot and Pod Necrosis

Alternaria leaf spot has been reported in all soybean-growing areas of the world. It occasionally appears in seedlings but generally attacks leaves and pods of plants approaching maturity. Because the disease usually occurs late in the growing season, yield losses are believed to be minimal. The use of broad-spectrum dithiocarbamate fungicides for control was reported in India.

An *Alternaria* sp. associated with pod necrosis was reported in 1987 in Canada.

Symptoms

Brown, necrotic leaf spots with concentric rings appear on infected foliage, enlarge, and may coalesce to form larger necrotic areas (Fig. 8). Infected leaves eventually dry out and drop prematurely. Seed infections increase with delays in harvesting.

Soybeans with chlorotic leaves and necrotic pods typically appear following periods of dry, warm weather in Ontario.

Causal Organisms

Alternaria spp. have been reported as the causal fungi of the leaf spot phase. *A. tenuissima* (Kunze ex Pers.) Wiltshire has been reported to cause a leaf spot in India and wilt in Kenya, and *A. atrans* Gibson is a weak parasite that invades leaves through aphid punctures and sunburn injury. *A. alternata* (Fr.) Keissler was consistently isolated from necrotic areas of pods in Canada.

Species of *Alternaria* are easily identified by their characteristic conidia, which are inverted club-shaped, mostly elongated at the tip, muriform at the bottom, and dark but usually lighter at the tip (Fig. 9). They are borne in long, simple chains and are mostly multiseptate.

The *Alternaria* spp. that cause leaf spot and are associated

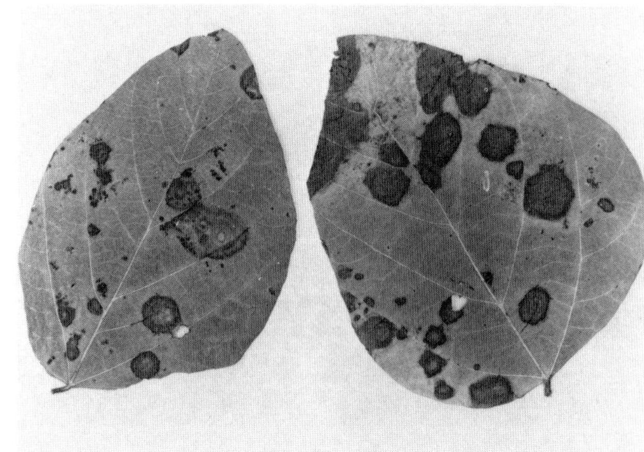

Fig. 8. Alternaria leaf spot, caused by an *Alternaria* sp., on soybean leaflets. (Courtesy D. W. Chamberlain)

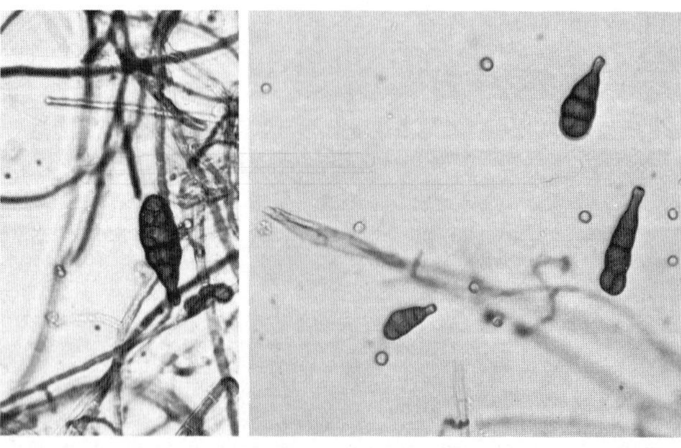

Fig. 9. Conidia of an *Alternaria* sp. as seen under a bright-field microscope. (Courtesy F. D. Tenne)

with pod necrosis are also important seedborne fungi (see Alternaria Pod and Seed Decay, under Seed Pathology).

Selected References

Anderson, T. R. 1987. Pod necrosis: A new disease affecting soybean on droughty soils in southwestern Ontario. (Abstr.) Annu. Meet. Can. Phytopathol. Soc., Ottawa, p. 142.

Mishra, B., and Prakash, O. 1975. Alternaria leaf spot of soybean from India. Indian J. Mycol. Plant Pathol. 5:95.

(Prepared by K. J. Cavanaugh)

Anthracnose

Soybean anthracnose, first reported in 1917 in Korea, occurs wherever soybeans are grown and causes damage in warm, humid areas. The disease reduces stands, seed quality, and yields by 16–26% or more in the United States, 30–50% in Thailand, and 100% in certain areas of Brazil and India. Losses increase as the percentage of infected pods rises. Infections limited to stems cause little yield reduction.

Symptoms

Soybeans are susceptible to anthracnose at all stages (Fig. 10). Symptoms typically appear in the early reproductive stages on stems, pods, and petioles as irregularly shaped brown areas and may resemble pod and stem blight (see Pod and Stem Blight and Phomopsis Seed Decay, under Fungal Diseases of Roots and Lower Stems). Stems, pods, and leaves may be infected without showing symptoms. In advanced stages of anthracnose, usually in the late reproductive stages, infected tissues are covered with black fruiting bodies (acervuli), which produce minute black spines (setae) that can be seen with the unaided eye (Figs. 11 and 12). The setae are diagnostic for the disease.

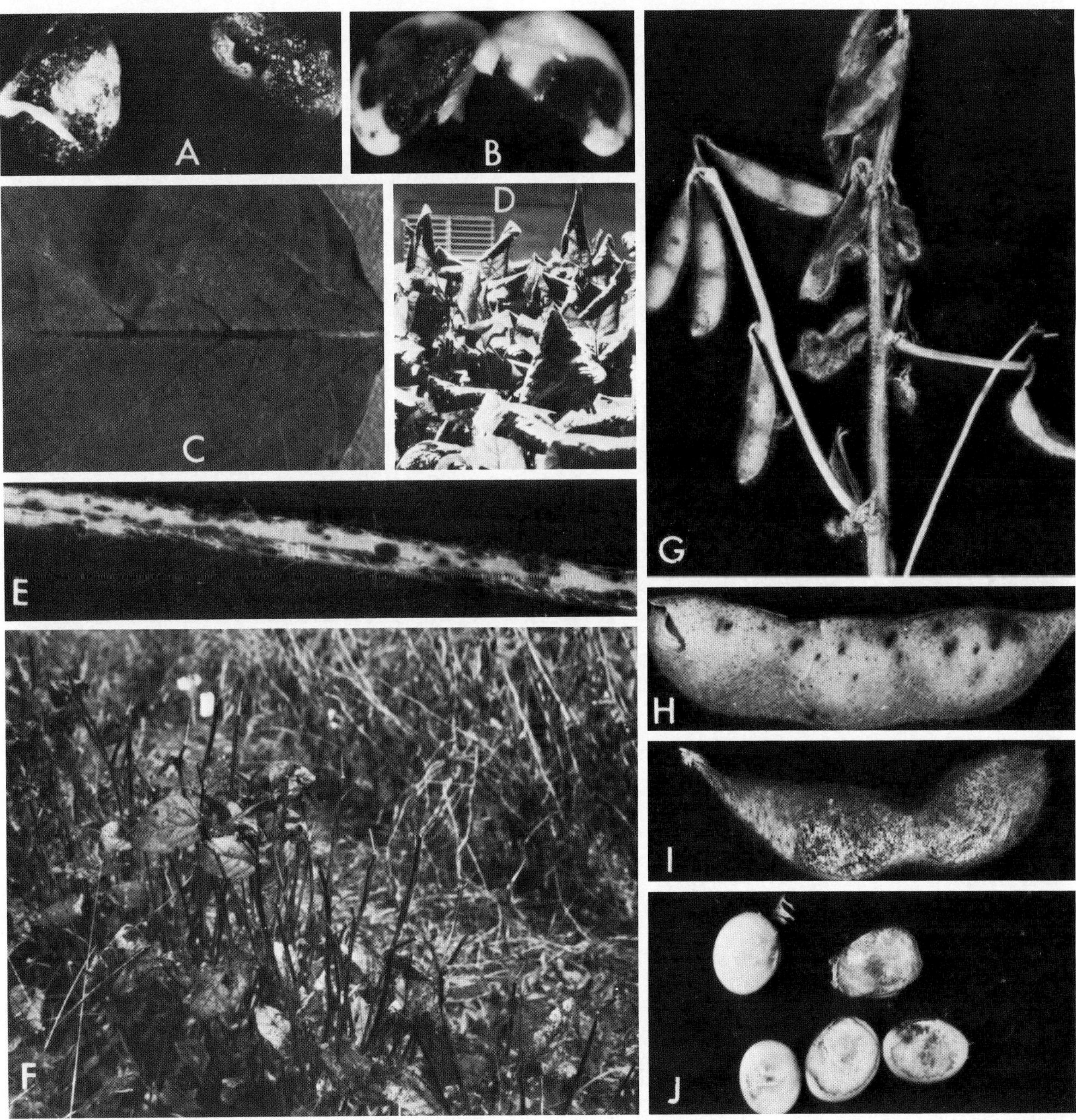

Fig. 10. Soybean anthracnose symptoms caused by *Colletotrichum truncatum*. **A,** Preemergence seed decay. **B,** Cotyledon cankers. **C,** Veinal necrosis of the lamina. **D,** Leaf rolling. **E,** Petiole cankers. **F,** Premature defoliation and blackening of petioles. **G,** Stem tip blight. **H,** Local lesions on a pod. **I,** Acervuli scattered on a pod. **J,** Brown staining of seeds. (Courtesy P. R. Hepperly and J. S. Mignucci)

Foliar symptoms that develop after prolonged periods of high humidity include necrosis of laminar veins, leaf rolling, petiole cankering, and premature defoliation (Fig. 10C–F). Blighted plants may be shorter than healthy plants.

Pre- and postemergence damping-off may occur when infected seeds are planted (Fig. 10A) (see *Colletotrichum truncatum*, under Seed Pathology). Dark brown, sunken cankers (lesions) often develop on the cotyledons of emerging seedlings (Fig. 10B). These cankers gradually extend up toward the epicotyl and down to the radicle. In humid weather, one or both cotyledons become water-soaked, quickly wither, and fall off. The anthracnose fungi may grow from infected cotyledons into young stems, where numerous small, deep-seated cankers form and may kill the young plants.

Anthracnose of maturing plants causes serious losses, particularly during rainy periods, when shaded lower branches and leaves are killed. Premature defoliation may occur throughout the canopy when cankers girdle the leaf petiole. Affected plants and whole fields senesce earlier than usual from the combined effects of stem and petiole lesions (Fig. 10G), and yields are reduced accordingly. When soybean pods or pedicels are infected early (Fig. 10H and I), either no seeds form (a condition known as pod blanking) or the seeds that develop are fewer or smaller than normal. Mycelium of the anthracnose fungi occasionally can completely fill the pod cavity, and seeds may become moldy, dark brown, and shriveled (Fig. 10J and Plate 5). Less severely infected seeds may show no visible sign of infection.

The infection may be asymptomatic, and symptoms and fruiting structures may be induced by treatment with paraquat.

Causal Organisms

The most common pathogen associated with soybean anthracnose is *Colletotrichum truncatum* (Schw.) Andrus & W. D. Moore (syns. *C. dematium* (Pers. ex Fr.) Grove var. *truncatum* (Schw.) Arx, *C. dematium* var. *truncata* (Schw.) Arx, and *C. glycines* Hori). Its teleomorph is unknown. Other *Colletotrichum* spp. that can be involved include *C. destructivum* O'Gara (teleomorph *Glomerella glycines* (Hori) Lehman & Wolf), *C. gloeosporioides* (Penz.) Sacc. (teleomorph

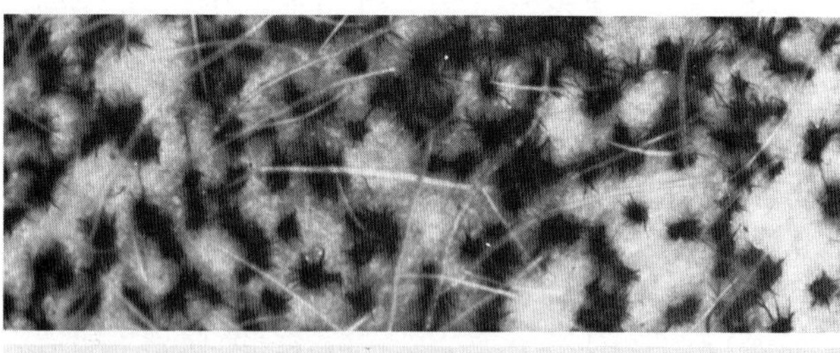

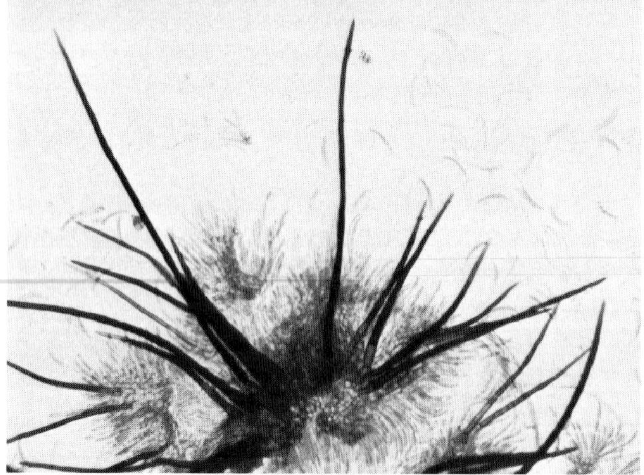

Fig. 11. Acervuli of the soybean anthracnose fungus *Colletotrichum truncatum* on a soybean stem (above); setae of an acervulus, with scattered conidia (below). (Courtesy C. C. Yeh)

G. cingulata (Ston.) Spauld. & Schrenk), and *C. graminicola* (Ces.) Wilson (teleomorph unknown). All have wide host ranges. All except *C. graminicola* are seedborne in soybeans.

C. truncatum is characterized by crowded, black acervuli borne on well-developed stromata. The acervuli are oval to elongate, hemispherical to truncate conical, and erumpent, with numerous black, needlelike, intermixed long and short setae, 60–300 × 3–8 µm (Fig. 11). Conidia, borne singly on conidiophores, are bluntly tapered, curved, unicellular, and hyaline and measure 17–31 × 3–4.5 µm (Fig. 12). They usually produce one to two short germ tubes. Germ tubes in contact with a solid surface produce dark, sticky appressoria, with direct penetration following.

C. truncatum grows well on oatmeal agar and potato-dextrose agar under alternating 12-hr periods of light and dark at an optimum temperature of 25°C. It produces whitish colonies that eventually turn smoky black with abundant acervuli.

G. glycines (anamorph *C. destructivum*) produces submerged perithecia that protrude through the epidermis from a stroma produced in the cortex. The perithecia form in groups of two or more and are short-beaked, membranous, globose, and 180–340 µm in diameter. Asci are oblong to bluntly clavate, 30–106 × 7–13.5 µm. Ascospores are ejected forcibly from perithecia upon maturation. Acervuli are black with numerous brown setae. Conidia are hyaline, one-celled, bluntly tapered, and straight and measure 20–22 × 4 µm. *G. glycines* forms acervuli and perithecia on potato-dextrose agar at 31–34°C. Isolates of this species and *C. truncatum* vary considerably in colony characteristics, size of fruiting structures, and pathogenicity. Germ tubes arising from conidia of both fungi develop large, black appressoria when in contact with the surface of a soybean plant or other solid surfaces. A narrow infection peg extends from the underside of the appressorium and directly penetrates the cuticle and cell wall. After penetration, hyphae spread both between and within host cells. In young, actively growing plants, the rate of spread is slow, and the fungus remains near the site of initial infection. As an infected plant reaches the flowering stage, the fungus resumes its growth and becomes systemic.

A key to the species found on soybeans grown in Illinois was developed by G. L. Hartman, J. B. Manandhar, and J. B. Sinclair:

1. Conidia straight (division I) 2
1. Conidia lunate (division II) 4

2. Conidia 5.6–6.2 µm wide (group I) .... *C. gloeosporioides*
2. Conidia less than 4.8 µm wide 3

3. Conidia 15–20.1 × 3.3–4.8 µm (group II) ...... *C. destructivum*
3. Conidia 16.3–17.8 × 2.8–3.9 µm (group III) .... *C. destructivum*

4. Conidia more than 24 µm long (group I) .... *C. truncatum*
4. Conidia less than 24 µm long (group II) 5

5. Conidia 21–24 µm long (group II) .... *C. truncatum*
5. Conidia less than 21 µm long (group III) .... *C. truncatum*

Disease Cycle and Epidemiology

C. truncatum and *G. glycines* can overseason as mycelium in infested crop residue or in infected seeds. Inoculum from infected seeds and debris may cause pre- and postemergence damping-off of seedlings. Mycelium may also become established in infected seedlings without symptoms developing until the plants begin to mature. Infected embryos fail to germinate. Stem and pod infections predominantly occur in the reproductive stages during warm, moist weather.

Plants are susceptible to infection by *C. truncatum* at all stages of development, particularly from bloom to pod fill, when yield-damaging infections typically occur. Disease forecasting methods used in the southeastern United States predict infections in this period and allow timely use of fungicides (see Soybean Disease Management Strategies). By

Fig. 12. Acervuli of *Colletotrichum truncatum*, the cause of soybean anthracnose, developing near a trichome, with setae and conidia, in a scanning electron micrograph (×250). (Courtesy N. M. Kokalis)

contrast, only mature plants (stage R_7) are susceptible to *G. glycines*.

Detailed information on environmental conditions that promote disease development is lacking. Conidia of the two fungi germinate and form appressoria at temperatures below 35°C when the plant surface is wet. The conidia are short-lived and sensitive to drying; 5 hr of air-drying can reduce germination by 98%. Infections typically occur if rain, dew, or fog provides free moisture for periods of 12 hr or more. There are no commercial resistant cultivars.

Control

1. Sow seeds relatively free of the pathogens.
2. Treat infected seeds with a recommended fungicide.
3. Plow under crop residues.
4. Spray with benomyl, fentin hydroxide, or chlorothalonil when conditions favoring infection occur between bloom and pod fill.
5. Rotate soybeans with other crops.

Selected References

Backman, P. A., Williams, J. C., and Crawford, M. A. 1982. Yield losses in soybeans from anthracnose caused by *Colletotrichum truncatum*. Plant Dis. 66:1032-1034.

Hartman, G. L., Manandhar, J. B., and Sinclair, J. B. 1986. Incidence of *Colletotrichum* spp. on soybeans and weeds in Illinois and pathogenicity of *Colletotrichum truncatum*. Plant Dis. 70:780-782.

Manandhar, J. B., Hartman, G. L., and Sinclair, J. B. 1986. *Colletotrichum destructivum*, the anamorph of *Glomerella glycines*. Phytopathology 76:282-285.

Manandhar, J. B., Hartman, G. L., and Sinclair, J. B. 1988. Soybean germ plasm evaluation for resistance to *Colletotrichum truncatum*. Plant Dis. 72:56-59.

Manandhar, J. B., Kunwar, I. K., Singh, T., Hartman, G. L., and Sinclair, J. B. 1985. Penetration and infection of soybean leaf tissues by *Colletotrichum truncatum* and *Glomerella glycines*. Phytopathology 75:704-708.

Brown Spot

Brown spot, also known as Septoria leaf spot, was reported in the United States in 1922 and first described in Japan in 1951. The disease has been reported in most southern and mid-Atlantic Coast states and in all North Central states, where it is most prevalent. It has also been reported in the cooler soybean-growing areas of Brazil, Canada, the People's Republic of China, Germany, Italy, Japan, Korea, the Soviet Union, Taiwan, and Yugoslavia.

Brown spot can cause premature defoliation. Yield losses have ranged from 8 to 15% under natural field infestation and from 8 to 34% in inoculated field plots. The disease is most severe when soybeans are grown continuously in the same field.

Symptoms

Brown spot is primarily a leaf spot disease, although seeds, stems, and pods of maturing plants are also infected. Irregular, dark brown spots varying from minute specks to areas 4 mm in diameter appear on both upper and lower surfaces of unifoliolate leaves. These leaves quickly turn yellow and drop. Adjacent lesions frequently coalesce to form irregularly shaped blotches. Numerous irregular, light brown lesions form on trifoliolate leaves (Fig. 13). These lesions gradually darken until they become chocolate brown to blackish brown. During warm, wet periods, the fungus progresses from lower to upper leaves. Late in the growing season, leaves become rusty brown and drop prematurely.

Irregular brown lesions with indefinite margins form on the main stem, branches, petioles, and pods. These lesions range from small specks to areas of several square centimeters. Symptoms on these organs are not sufficiently distinct from those of other diseases to be diagnostic.

Causal Organism

The brown pycnidia of *Septoria glycines* Hemmi (teleomorph *Mycosphaerella uspenskajae* Mashk. & Tomil.), the causal organism, form in dead tissues of old lesions. The pycnidia are embedded in the substrate and open to the surface by a large pore (ostiole). Pycnidia formed in leaf tissues are globose to conical-globose and generally open to the upper surface. Those formed in stems are considerably flattened because of their position in the tissue. The pycnidia are 60–125 μm (mostly 90–100 μm) in diameter and are scattered to densely packed. The pycnidial wall is thin and membranous. Conidia (pycnidiospores) are hyaline, filiform, and curved, 21–50 × 1.4–2 μm and generally 35–40 μm long. They are indistinctly one- to three-septate, with septation becoming noticeable at the time of germination. The conidia readily germinate in water on the surface of leaves. First, germ tubes extend from the ends of the spores. Later, hyphae develop from other conidial cells. Mature hyphae are densely branched, thick-walled, and

Fig. 13. Leaf lesions caused by the brown spot fungus, *Septoria glycines*, varying from minute specks (on the upper part of the leaflet) to larger areas (on the lower part). (Courtesy R. F. Nyvall)

beadlike. Pathogenic variability among isolates of the fungus has not been detected.

Disease Cycle and Epidemiology

Primary inoculum arises from conidia and mycelium overseasoning on infected leaf and stem debris and in diseased seeds. Lesions that develop on infected cotyledons and unifoliolate leaves are sources of inoculum for later infections.

Infection and disease development are favored by warm, moist weather, which promotes sporulation of the pathogen on primary lesions. Conidia are distributed by wind and splashing rain. Infection and defoliation proceed from the lower to the upper parts of the plant. Dry weather halts the spread of the disease.

The fungus enters through stomata and grows intercellularly, killing cells next to the hyphae. It penetrates pods through stomata or by growing through placental and funicular tissue and later invades the seed coat.

Sources of resistance in soybean have not been found in over 10,000 accessions screened. Some wild species of *Glycine* appear to be resistant.

Control

1. Plow under all crop residues after harvest.
2. Plant seeds relatively free of the pathogen.
3. Use a less susceptible cultivar. Differences in susceptibility among cultivars have been observed, and resistant cultivars may become available.
4. Rotate soybeans with a crop that is not susceptible to brown spot, for at least one year.
5. Apply a foliar fungicide from bloom to pod fill.

Selected References

Hemmi, T. 1915. A new brown-spot disease of the leaf of *Glycine hispida* Maxim. caused by *Septoria glycines* sp. (In Japanese) Trans. Sapporo Nat. Hist. Soc. 6:23-27.

Kamicker, T. A., and Lim, S. M. 1985. Field evaluation of pathogenic variability in isolates of *Septoria glycines*. Plant Dis. 69:744-746.

Lim, S. M. 1980. Brown spot severity and yield reduction in soybean. Phytopathology 70:974-977.

Lim, S. M. 1983. Responses to *Septoria glycines* of soybeans nearly isogenic except for seed color. Phytopathology 73:719-722.

Lim, S. M., and Hymowitz, T. 1987. Reactions of perennial wild species of genus *Glycine* to *Septoria glycines*. Plant Dis. 71:891-893.

MacNeill, B. H., and Zalasky, H. 1957. Histological study of host-parasite relationships between *Septoria glycines* Hemmi and soybean leaves and pods. Can. J. Bot. 35:501-505.

Williams, D. J., and Nyvall, R. F. 1980. Leaf infection and yield losses caused by brown spot and bacterial blight diseases of soybeans. Phytopathology 70:900-902.

Wolf, F. A., and Lehman, S. G. 1926. Brown spot disease of soybeans. J. Agric. Res. 33:365-374.

Young, L. D., and Ross, J. P. 1979. Brown spot development and yield response of soybean inoculated with *Septoria glycines* at various growth stages. Phytopathology 69:8-11.

(Prepared by S. M. Lim)

Cercospora Blight and Leaf Spot

Cercospora blight and leaf spot have been reported in Brazil, Japan, Taiwan, Uganda, and most soybean-growing regions in the United States. Losses from the disease in the United States were estimated at 130.5 million bushels (3.5 million metric tons) in 15 southern states in 1978 and 18.4 million bushels (0.5 million metric tons) in 1979. Losses from the disease in Brazil range from 30% in the north to 15% in the central region. (See also Purple Seed Stain, under Seed Pathology.)

Symptoms

The first symptoms of the disease are observed at the beginning of and throughout seed set. Upper leaves exposed to the sun have a light purple appearance. The discoloration can deepen and extend over the entire upper surface of affected leaves, giving the leaves a leathery, dark, reddish purple appearance highlighted with bronzing. Reddish purple, angular to irregular lesions later occur on both the upper and the lower surfaces of leaves (Fig. 14). The lesions vary from pinpoint spots to irregular patches up to 1 cm in diameter, which may coalesce to form large necrotic areas. Veinal necrosis may also be observed. Numerous infections cause rapid chlorosis and necrosis of leaf tissues, resulting in defoliation starting with the upper leaves. These symptoms are often mistaken for early senescence; however, green leaves usually remain below the defoliated area.

Lesions on petioles and stems are slightly sunken, reddish purple areas up to several millimeters long. Infection of petioles increases defoliation, with the petioles remaining attached to plants. The most obvious symptom is the blighting of young, upper leaves over large areas, even entire fields. On more susceptible cultivars, round, reddish purple lesions, which later become purplish black, occur on pods. A general dark reddish purple discoloration, similar to that of upper leaves, also occurs on upper stems, petioles, and pods exposed to the sun. Infection may be latent, with no symptoms expressed.

Causal Organism

For details concerning the causal fungus, *Cercospora kikuchii* (T. Matsu. & Tomoyasu) Gardner, see Purple Seed Stain, under Seed Pathology. Isolates of *C. kikuchii* that produce high levels of foliar blight differ in their levels of seed infection. Some isolates from soybean seeds do not cause leaf blight symptoms.

Fig. 14. Symptoms caused by *Cercospora kikuchii* on a soybean leaflet. (Courtesy C. C. Yeh)

Disease Cycle and Epidemiology

Primary inoculum may come from infected seeds or surface debris from the previous crop. Secondary inoculum comes from infected plant tissues. Most soybean fields surveyed in Arkansas during the full-pod growth stage had some damage from the disease. In the field, temperatures of 28–30°C, along with extended periods of high humidity, favor disease development. In a dew chamber, maximum infection occurred under conditions of alternating 12-hr periods of light and dark and between 20 and 24°C. Infection increased with the duration of the dew period. The disease is more severe in early-maturing cultivars.

Control

The soybean cultivars Bedford, Bragg, Dare, Forrest, Hood 75, Mack, and Pickett 71 are susceptible to Cercospora leaf blight; Davis and Lee 74 are moderately susceptible; and Tracy is slightly susceptible. Hood 75 is resistant to purple seed stain but susceptible to the leaf blight phase.

Selected References

Dakai, Y. 1984. Importance of fallen soybean leaves infected with *Cercospora kikuchii* Matsu. & Tomoyasu for disease spread. (In Japanese) Bull. Hiroshima Prefectural Agric. Exp. Stn. 48:67-74.

Martin, K. F., and Walters, H. J. 1982. Infection of soybean by *Cercospora kikuchii* as affected by dew temperature and duration of dew periods. (Abstr.) Phytopathology 72:974.

Sanders, R. L., and Abney, T. S. 1985. Soybean leaf blight and seed infection caused by isolates of *Cercospora kikuchii* of diverse origin. (Abstr.) Phytopathology 75:966.

Walters, H. J. 1980. Soybean leaf blight caused by *Cercospora kikuchii*. Plant Dis. 64:961-962.

(Prepared by J. C. Rupe)

Choanephora Leaf Blight

Choanephora leaf blight has been reported in tropical and subtropical areas of Thailand and the United States. Yield losses of 5–10% and 20–40% have been reported in northern and central Thailand, respectively. Other reported hosts include cowpea (*Vigna unguiculata* (L.) Walpers), cucumber (*Cucumis sativus* L.), mung bean (*V. radiata* (L.) R. Wilczek), and tobacco (*Nicotiana* sp.).

Symptoms

Symptoms occur in the field after two or three consecutive days of rain when the weather is cool and the relative humidity is close to saturation. The first symptom of the disease appears on mature leaves and occasionally on young leaves at the top of the plant. Pods may also be infected. Infected areas of leaves first appear grayish, similar to leaves with hot-water scald injury, and then turn dark. If high humidity and cool weather prevail at night, sporangiophores and sporangia form, which can be easily seen by the unaided eye on infected plant parts early in the morning (Fig. 15). Infected leaves may drop and adhere to the stems beneath. When conditions favorable for the disease continue, extensive defoliation occurs, which results in yield loss.

Causal Organism

The causal fungus is *Choanephora infundibulifera* (Currey) Sacc. (syn. *C. trispora* Thaxter). No perfect stage has been described. Sporangiophores bearing sporangiola arise from aerial hyphae, which are erect or ascending, up to 10 mm high, unbranched, hyaline, smooth-walled, and up to 30 μm wide. They are apically dilated to form a clavate vesicle, up to 45 μm in diameter, from which arise one to six unbranched or dichotomously branched, distally clavate secondary vesicles, which bear the sporangiola on apical dilations. As many as 12 or more sporangioliferous dilations are formed. Secondary vesicles usually remain attached at maturity. Sporangiospores within sporangiola are broadly ellipsoidal to obovoid, brown or reddish brown to pale brown, and smooth or, rarely, indistinctly longitudinally striate. They remain inside the wall of the sporangiolum at maturity. Colonies grow rapidly on potato-carrot agar at 25°C.

Control

1. Remove and destroy infected plant parts to prevent secondary spread of the disease.

2. Use resistant cultivars where available.

3. No satisfactory chemical control has been reported, although triforine (Saprol, 18.2%) and copper oxychloride were proven effective in laboratory studies.

Selected References

Kirk, P. M. 1984. A Monograph of the Choanephoraceae. Mycol. Pap. 152. Commonwealth Mycological Institute, Kew, Surrey, England.

Poonpolgul, S., and Pupipat, U. 1978. Soybean diseases of Thailand. (In Thai; English summary) Kasetsart Univ. J. 12:143-154.

Sirivallop, P., Visutharom, T., Boonthavee, S., and Wannapee, L. 1986. Effectiveness of fungicide for controlling Choanephora wet rot. (In Thai; English summary) Agric. Dep. Annu. Rep., Bangkok, pp. 858-867.

(Prepared by C. Sittigul and N. Suwarnarat)

Downy Mildew

Downy mildew, one of the most consistently occurring foliar diseases of soybean, was first reported in the United States in 1923. It is distributed worldwide but is seldom a serious threat to soybean production. It causes some defoliation, reduces seed quality and size, and may lower yields by as much as 8%.

Symptoms

Downy mildew appears on the upper surface of young leaves as pale green to light yellow spots, which enlarge into pale to bright yellow lesions of indefinite size and shape, depending on leaf age (Fig. 16). They may later turn grayish brown to dark brown with a yellowish green margin and may finally become entirely brown. On the lower surface of leaves, particularly in moist weather, lesions are covered with tufts of grayish to pale purple sporangiophores, which easily distinguish downy mildew from other foliar diseases of soybeans. Severely infected

Fig. 15. Symptoms caused by *Choanephora infundibulifera* on a soybean leaflet. (Courtesy S. P. Poonpolgul)

leaves turn yellow and then finally brown, curl at the edges, and drop prematurely.

Pod infections may occur without external symptoms. The interior of pods and the seed coat may be encrusted with a whitish mass of mycelia and oospores. Seeds partly or completely encrusted with oospores often appear dull white and have cracks in the seed coat (Fig. 17). They may be smaller or lighter in weight than normal seeds. (See *Peronospora manshurica*, under Seed Pathology.)

Infected seeds can produce systemically infected seedlings. Symptoms first appear when the plants are about 2 weeks old. Light green areas appear at the base of primary leaves and spread along veins in a serrated or fanlike manner. Not all trifoliolate leaves show symptoms. Systemically infected plants may remain small, with stunted, mottled, gray-green leaves that curl down at the edges. Prolific sporangiophores can occur on underleaf surfaces.

Causal Organism

The hyphae of *Peronospora manshurica* (Naum.) Syd. ex Gäum. (syn. *P. sojae* Lehman & Wolf), the causal fungus of downy mildew, are intercellular in host tissues, coenocytic, and 7–10 μm wide. Sporangiophores are slender (240–984 × 5–9 μm) and gray to pale violet, and they branch dichotomously two to

Fig. 16. Downy mildew symptoms, caused by *Peronospora manshurica*, on the lower surface (left) and upper surface (right) of soybean leaflets. (Courtesy D. W. Chamberlain)

Fig. 17. Soybean seed encrusted with oospores of *Peronospora manshurica*, the downy mildew fungus. (Courtesy U.S. Department of Agriculture)

10 times (Fig. 18). They emerge singly or in clusters from stomata on the lower surface of leaves. The terminal branchlets, or sterigmata (9–13 × 2–3 μm), are pointed and straight. Sporangia (usually 19–24 μm in diameter) are subhyaline and broadly elliptical to subglobose (Fig. 18). They lack an apical papilla and germinate in water from an indefinite point.

As in other species of *Peronospora*, antheridia fertilize oogonia. The resulting light brown or yellow oospores may be smooth or variously marked and have a reticulated wall (Fig. 18). Including the reticulations, which are more visible if stained with cotton blue, the oospores measure 20–45 μm in diameter. They germinate and form a germ tube.

The viability of oospores can be determined by incubating them in 1 ml of distilled water for 48 hr at 30°C and then adding one drop of a 1% solution of 2,3,5-triphenyltetrazolium chloride; after 48 hr, viable spores turn orange-red.

Intercellular mycelia form haustoria, which invade adjacent cells. The haustoria vary in diameter from 1 to 3.5 μm and are generally long, cylindrical, tenuous, and curving. They branch up to five times within a cell.

Thirty-three physiologic races of the pathogen have been identified in the United States on the basis of disease reactions of differential cultivars. Cultivars with the resistance gene *Rpm* are resistant to races 1 through 32 but are susceptible to race 33. However, some cultivars are resistant to this race.

Disease Cycle and Epidemiology

P. manshurica overseasons as oospores in leaves and on seeds. Planting oospore-encrusted seeds may produce a few systemically infected seedlings under cool conditions.

Seedling infection probably occurs in the hypocotyl from oospores on the seed coat or leaf residues. Hyphae grow through the hypocotyl into the first two or three trifoliolate

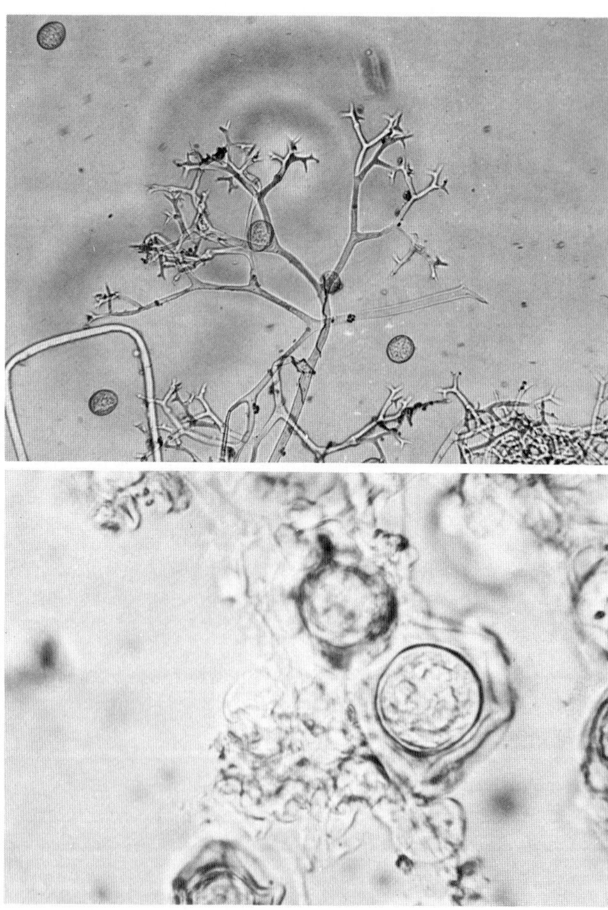

Fig. 18. The downy mildew fungus, *Peronospora manshurica*: sporangiophores with sporangia (above) and mature oospores (below), in bright-field photomicrographs. (Courtesy F. D. Tenne [above] and H. C. Phatak [below])

18

leaves and into some nodal buds. The first and second pairs of leaves to develop on systemically infected plants show large chlorotic areas. When dew is present, sporangia are produced on newly infected leaves and are disseminated by air currents. Few or no sporangia form on old leaves. Typically, sporangia germinate within 12 hr.

The germ tube may enter a stoma or form an ovate appressorium, about $9 \times 12 \mu$m. Penetration pegs push between adjacent cell walls into the intercellular spaces of the mesophyll. In susceptible soybean cultivars, the fungus continues to ramify through the mesophyll and into the palisade layers. Haustoria are frequently formed in the cells. Bundle sheath parenchyma cells of the larger veins stop hyphal advancement, with the result that the lesions acquire an angular shape. In resistant cultivars, hyphal growth stops abruptly, and relatively few haustoria are formed.

Downy mildew is favored by high humidity and temperatures of 20–22°C. Sporulation occurs between 10 and 25°C; no sporulation occurs above 30°C or below 10°C.

Leaves become more resistant to local infection as they grow older. A soybean cultivar may appear to be highly susceptible or highly resistant, depending on leaf age at the time of infection. Leaves inoculated when 5 or 6 days old give a susceptible reaction, whereas leaves 8 days old or older give a more resistant reaction. The number of lesions increases and their size decreases as leaves age. Leaves of plants exposed to high temperature before infection also have a more resistant reaction.

Control

1. Treat seeds with a fungicide.
2. Plow under soybean residue.
3. Rotate soybeans with a crop not susceptible to downy mildew, for one year or more.
4. Grow resistant cultivars when feasible.

Selected References

Dunleavy, J. M. 1987. Yield reduction in soybeans caused by downy mildew. Plant Dis. 71:1112-1114.

Inaba, T., and Morinaka, T. 1983. Effect of environmental factors on oospore production of soybean downy mildew fungus, *Peronospora manshurica*. (In Japanese) Bull. Natl. Inst. Agric. Sci. (Jpn.) 38:121-130.

Kivik, N. N., Koshevskii, I. I., and Hong, N. S. 1985. Effect of ecological conditions on the biology of *Peronospora manshurica* (Naum.) Syd. (In Russian) Mikol. Fitopatol. 19:422-426.

Lim, S. M., Bernard, R. L., Nickell, C. D., and Gray, L. E. 1984. New physiological race of *Peronospora manshurica* virulent to the gene *Rpm* in soybeans. Plant Dis. 68:71-72.

Pathak, V. K., Mathur, S. R., and Neergaard, P. 1978. Detection of *Peronospora manshurica* (Naum.) Syd. in seeds of soybean, *Glycine max*. EPPO Bull. 8:21-28.

Riggle, J. H., and Dunleavy, J. M. 1974. Histology of leaf infection of susceptible and resistant soybeans by *Peronospora manshurica*. Phytopathology 64:522-526.

(Prepared by D. V. Phillips)

Frogeye Leaf Spot

Frogeye leaf spot, sometimes called Cercospora leaf spot, was first reported in soybeans in 1915 in Japan and was first observed in the United States in 1924. The disease occurs worldwide but is most common in warmer regions during warm, humid weather. Use of resistant cultivars has reduced the incidence of the disease and precluded major economic losses. Yields from susceptible cultivars may be reduced by 15%.

Symptoms

Frogeye leaf spot is primarily a disease of foliage, but stems, pods, and seeds may also be infected. The lesions are circular to

angular spots varying in size from less than 1 mm to 5 mm in diameter. When they are first visible on the upper surface of leaves, they are approximately the size of mature lesions. They begin as dark, water-soaked spots, with or without a lighter center, and develop into brown spots surrounded by a narrow, dark reddish brown margin (Fig. 19). As the lesions age, the central area becomes ash gray to light brown. On the lower surface of leaves, the spots are darker, having light to dark gray centers while sporulating, with clusters of conidiophores, and a thin, reddish brown margin. The nonsporulating older lesions are light to dark brown and translucent and often have a paper white center containing minute dark stromata. Several of these lesions may coalesce to form larger, irregular spots. When lesions are numerous, leaves wither and fall prematurely.

Stem infections appear late in the season and are less common and distinctive than foliar infections. The lesions are two to four times as long as they are wide and may extend up to half the circumference of the stem. Young lesions are deep red with a narrow, dark brown to black margin; the central area is flattened or slightly sunken. As they age, their centers become brown and then pale smoky gray. Minute, dark stromata, often bearing clusters of conidiophores and numerous conidia, form in older lesions, giving them an almost black appearance.

Lesions on pods are circular to elongate, slightly sunken, and reddish brown (Fig. 20). Older lesions become brown to light gray, usually with a narrow, dark brown border. The causal fungus frequently grows through pod walls into maturing seeds (see *Cercospora sojina*, under Seed Pathology). Papillate or

Fig. 19. Frogeye leaf spot, caused by *Cercospora sojina*, on the upper surface (left) and lower surface (right) of soybean leaflets. (Courtesy J. T. Yorinori)

Fig. 20. Pod lesions caused by the frogeye leaf spot fungus, *Cercospora sojina*. (Courtesy J. T. Yorinori)

elevated conical lesions develop where the infected membranous inner lining of the pod is in contact with the seeds.

Causal Organism

The frogeye leaf spot fungus, *Cercospora sojina* Hara (syn. *C. daizu* Miura), is highly variable. Conidiophores arise in fascicles of two to 25 from a thin stroma (Fig. 21). They are light to dark brown and are 52–120 × 4–6 μm. They have one to several septations with prominent bands (geniculations) and spore scars (Fig. 22). Conidia are borne on the tips of the conidiophores and are pushed aside as the conidiophores continue their indeterminate growth (Fig. 21). Generally one to three but sometimes up to 11 conidia are formed on a single conidiophore.

The conidia are zero- to 10-septate, hyaline when young, and

Fig. 21. Conidiophores of *Cercospora sojina*, the cause of frogeye leaf spot, arising in a fascicle, with detached conidia on a leaf surface, in a scanning electron micrograph (×450). (Courtesy N. M. Kokalis)

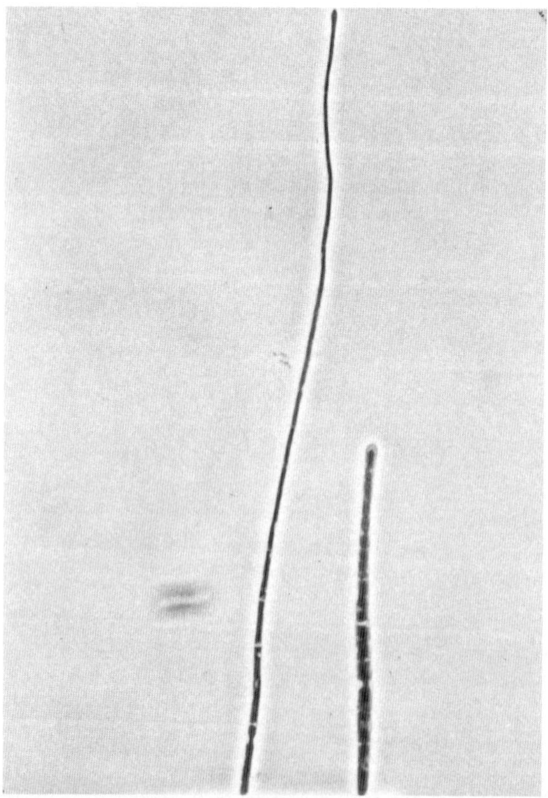

Fig. 22. Conidia of *Cercospora kikuchii*, the cause of purple seed stain and Cercospora leaf spot (left), and *C. sojina*, the cause of frogeye leaf spot (right). (Courtesy J. T. Yorinori)

elongate to fusiform, tapering toward the tip (Fig. 22). The base of a conidium is usually rounded and bears circular scars marking the attachment to the conidiophore. On infected leaves, conidia are 24–108 × 3–9 μm; a size range of 40–60 × 6–8 μm is most common. The size and shape of the conidia and conidiophores vary with the substrate on which the fungus grows. The conidia of *C. sojina* are much shorter than those of *C. kikuchii* (Fig. 22).

Conidia begin to germinate within an hour in water, producing one to several germ tubes from the end cells and occasionally from the intervening cells. In culture, they frequently germinate in situ, forming short germ tubes that may in turn produce secondary conidia. Few conidia germinate after 6 months at room temperature, but they have been reported to withstand longer periods of drying.

Some *C. sojina* isolates sporulate poorly on most media. Relatively abundantly sporulating isolates sporulate best on V-8 juice agar or lima bean agar. On water agar, sparse aerial hyphae develop in about 1 week, followed by the formation of smoky brown conidiophores with conidia. Colonies develop on potato-dextrose agar as compact mycelial mats in alternating light and dark bands, which give the appearance of concentric rings. The surface is convoluted or folded radially. Seeds infested with *C. sojina* can be identified by plating seeds that have not been surface-disinfected on moist blotters and examining them with a stereoscope.

Five physiologic races of the fungus have been reported in the United States. A set of differential cultivars to separate these races is presented in Table 5. Race 2 is seedborne but does not sporulate on soybean seeds.

Disease Cycle and Epidemiology

The fungus survives as mycelium in infected seeds and infested soybean debris. Germination of infected seeds is slightly reduced. Infected seeds that germinate may produce weak, stunted seedlings with lesions on the cotyledons. Sporulation on the cotyledon lesions provides inoculum for infecting young leaves. On inoculated plants the first lesions are visible after 9 to 12 days, and the first spores are produced within the next 24 to 48 hr. In warm, humid weather, sporulation is profuse in the resulting lesions. Conidia are carried short distances by air currents and splashing rain, and under favorable conditions secondary infections of leaves, stems, and pods arise throughout the season. Young leaves are infected more rapidly than older leaves; with adequate moisture, new leaves become infected as they develop, until often all the leaves of a plant are infected. Leaves that develop and become fully expanded during dry periods unfavorable for invasion may remain relatively disease-free. As a result, a layered pattern of heavily diseased and very lightly diseased leaves sometimes occurs on the same plant. At first, only seeds directly in contact with pod lesions become discolored, but discoloration may extend to other seeds as the pod matures.

Reactions of different cultivars to *C. sojina* vary from immunity to high susceptibility. The more susceptible the cultivar, the more uniform and larger the spots, with profuse sporulation. Cultivars having an intermediate reaction have fewer sporulating spots, most of which are less than 2 mm in diameter. Some cultivars with an intermediate reaction may not show infection when subjected to inoculation by conidia

Table 5. Reaction of Selected Soybean Cultivars to Races 1–5 of *Cercospora sojina*

Cultivar	Race of *Cercospora sojina*[a]				
	1	2	3	4	5
Blackhawk	S	S	S	S	S
Lincoln	R	S	R	U	R
Lee	R	R	S	S	R
Hood	R	R	R	S	S
Davis	R	R	R	R	R

[a]S = susceptible; R = resistant; U = reaction unknown.

naturally produced on nearby susceptible plants but may be infected when artificially inoculated, either in the field or in the greenhouse. This is an important aspect considered in selection for resistance to *C. sojina* in Brazil. Many breeding lines with high yield potential and desirable agronomic traits may be unnecessarily discarded when selection is based solely on the presence or absence of spots following artificial inoculation.

The single dominant genes Rcs_1, Rcs_2, and Rcs_3, controlling resistance to races 1, 2, and 5, respectively, have been described. Rcs_3 also conditions resistance to race 2 and perhaps other races, since it is found in the cultivar Davis, which is resistant to all known races. Some cultivars resistant to races 1 and 5 have been heavily infected in the southern United States and Brazil; this indicates the presence of additional races.

Control

1. Grow adapted, resistant cultivars.
2. Plant high-quality seeds relatively free of the pathogen.
3. Rotate soybeans with other crops for two years.
4. Treat seeds with a fungicide.
5. Apply fungicides at growth stages R_2-R_5.
6. Plow under crop residues.

Selected References

Athow, K. L. 1973. Fungal diseases. Pages 461-462 in: Soybeans: Improvement, Production, and Uses. B. E. Caldwell, ed. American Society of Agronomy, Madison, WI.
Athow, K. L., Probst, A. H., Kurtzman, C. P., and Laviolette, F. A. 1962. A newly identified physiological race of *Cercospora sojina* on soybean. Phytopathology 52:712-714.
Backman, P. A., Rodríguez-Kábana, R., Hammond, J. M., and Thurlow, D. L. 1979. Cultivar, environment, and fungicide effects on foliar disease losses in soybeans. Phytopathology 69:562-564.
Boerma, H. R., and Phillips, D. V. 1983. Genetic implications of the susceptibility of Kent soybean to *Cercospora sojina*. Phytopathology 73:1666-1668.
Lehman, S. G. 1928. Frog-eye leaf spot of soybean caused by *Cercospora diazu* Miura. J. Agric. Res. 36:811-833.
Phillips, D. V., and Boerma, H. R. 1982. Two genes for resistance to race 5 of *Cercospora sojina* in soybeans. Phytopathology 72:764-766.

(Prepared by D. V. Phillips and J. T. Yorinori)

Phyllosticta Leaf Spot

Phyllosticta leaf spot, or Phyllosticta leaf blight, was reported in Italy in 1900 and in the United States in 1927. Since then, it has been observed in North Borneo, Bulgaria, Canada, the People's Republic of China, Germany, Japan, and the Soviet Union. It occurs throughout the soybean belt in the United States. Cool, moist conditions favor disease development. The disease may cause severe defoliation early in the season but is generally considered of minor importance.

Symptoms

Plants of all ages are affected. Lesions form primarily on leaves. Leaf spots are round to oval, irregular, or V-shaped and up to 2 cm in diameter (Fig. 23). Usually only the first few trifoliolate leaves are affected. In young plants, infection starts at the leaf margin and progresses inward, forming an irregular, V-shaped area. The spots, at first pale green, become dull gray to tan, with a narrow, dark brown or purplish border. Numerous small, black specks (pycnidia) form in older lesions.

The fungus may progress from the leaf blades into the petioles and then to the stipules and stem tissues at the leaf scar. Superficial lesions 1-4 mm wide and up to 12 mm long sometimes form on petioles, stems, and pods. These lesions are light gray, tan, or brownish with a narrow brown or purplish border. Lesions on pods are circular with a reddish margin. Seeds beneath pod lesions may become infected.

Causal Organism

The causal fungus is *Phyllosticta sojicola* Massal. (syn. *P. glycinea* Tehon & Daniels); its teleomorph is *Pleosphaerulina sojicola* (Massal.) Miura. The pathogen has amphiphyllous, globose pycnidia, which resemble those of *Ascochyta*; they are 65–170 μm in diameter, with an ostiole 10–20 μm across. Conidia ($1.8-3.2 \times 5.2-7$ μm) are one-celled, narrowly elliptical with rounded ends, and hyaline, with inconspicuous guttulae. The pathogen is seedborne in soybeans.

Control

1. Plant seeds relatively free of the pathogen.
2. Rotate soybeans with other crops.
3. Plow under crop residues.

Selected References

Crall, J. M. 1948. Defoliation of soybeans in southeast Missouri caused by *Phyllosticta glycineum*. Plant Dis. Rep. 32:184-186.
Jehle, R. A., Jenkins, A. E., Kreitlow, K. W., and Sherwin, H. S. 1952. An outbreak of Phyllosticta canker and leaf spot of soybeans in Maryland. Plant Dis. Rep. 36:155-158.
Olive, L. S., and Walker, E. A. 1944. A severe leafspot of soybean caused by *Phyllosticta sojaecola*. Plant Dis. Rep. 28:1122-1123.
Walters, H. J., and Martin, K. F. 1981. *Phyllosticta sojaecola* on pods of soybeans in Arkansas. Plant Dis. 65:161-162.

Powdery Mildew

Powdery mildew, first reported in Germany in 1921, has since been reported in Brazil, Canada, the People's Republic of China, India, Puerto Rico, South Africa, and the United States. The disease is common in greenhouse-grown plants and was reported in many fields in the southeastern and midwestern United States during the 1970s. In 1974, severe outbreaks were reported in Delaware, Georgia, Illinois, Iowa, and Wisconsin. Yield losses of 10–35% have been recorded in susceptible cultivars.

The causal fungus also infects bean (*Phaseolus vulgaris* L.), mung bean (*Vigna radiata* (L.) R. Wilczek), pea (*Pisum sativum* L.), cowpea (*V. unguiculata* (L.) Walpers), and some wild soybeans (*Glycine* spp.), as well as some species of the Caprifoliaceae and Solanaceae.

Symptoms

White, powderlike patches composed of mycelia and conidia form on cotyledons, stems, pods, and leaf surfaces and merge to cover all aboveground plant parts (Fig. 24). Different cultivars

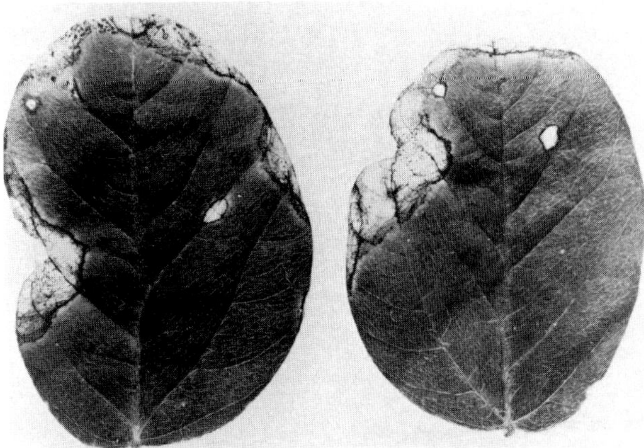

Fig 23. Leaf spots and lesions caused by *Phyllosticta sojicola*. (Courtesy U.S. Department of Agriculture)

may develop different symptoms—chlorosis, green islands, rusty patches, defoliation, or combinations of these; alternatively, the fungus may grow luxuriantly without producing visible symptoms. Some cultivars show susceptible reactions as seedlings but demonstrate resistance as mature plants. Chlorotic spots and veinal necrosis indicate a hypersensitive reaction. Powdery mildew colonies may become darkened when parasitized by *Cicinnobolus cesati* d By. (syn. *Ampelomyces* sp.).

Causal Organism

The causal fungus is *Microsphaera diffusa* Cke. & Pk. (syns. *Erysiphe polygoni* DC., *E. glycines* Tai, and *Microsphaera* sp.). Mature cleistothecia and their appendages are needed for positive identification of the pathogen (Fig. 25); the appendages of immature cleistothecia can lead to misidentification. Cleistothecia are only rarely produced in most areas. They develop abundantly in blue lupine (*Lupinus*), however. Immature cleistothecia, before the development of appendages, are light yellow to tan. As maturity approaches, they darken and turn rusty brown, and unbranched appendages begin to develop. Mature cleistothecia (55–126 μm in diameter) are hemispherical and black and have four to 50 appendages, the length of which is 1.5–7 times the diameter of the cleistothecium (Fig. 25). The appendages are thick-walled, hyaline to pale brown at the base, and dichotomously or subdichotomously branched three to five times at the apex. Multiple pyriform asci with short attachment stalks are released through broken cleistothecial walls. Each ascus contains up to six yellow, ovoid ascospores, measuring 9×18 μm.

M. diffusa is an obligate parasite that produces mycelia and conidia on plant surfaces and intracellular haustoria in epidermal cells. Large numbers of conidia (27.7–54.1 $\times$ 17.1–21.1 μm) are produced in chains borne on short, simple conidiophores (Fig. 26). The conidia are slightly barrel-shaped, with flattened ends. Dispersal of the fungus, rapid colonization, and spread of the disease in the field are highly dependent on the asexual stage.

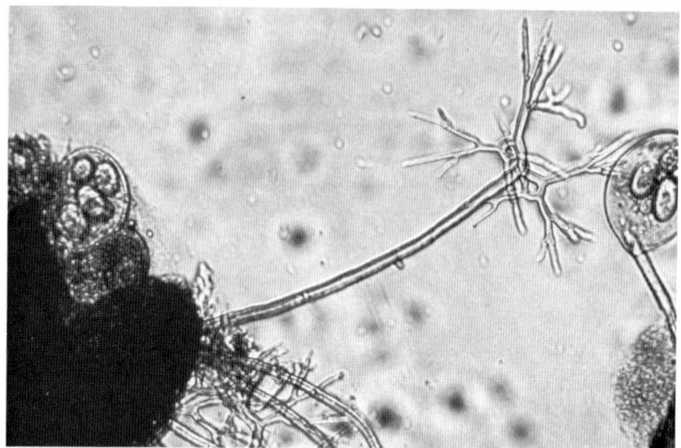

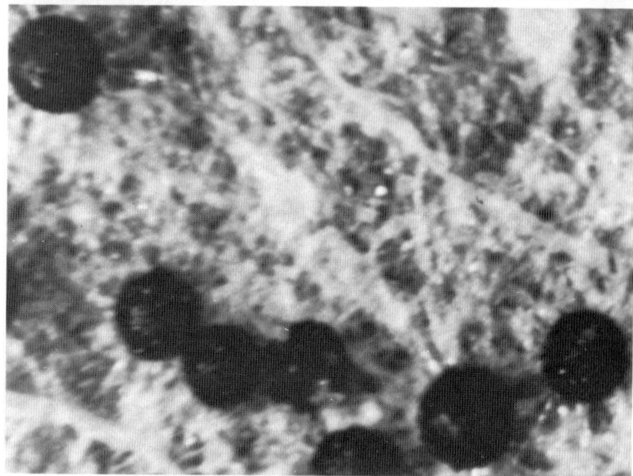

Fig. 25. *Microsphaera diffusa*, the causal fungus of powdery mildew: an open cleistothecium with an escaping ascus, containing ascospores, and a dichotomously branched appendage (above); a cluster of cleistothecia and mycelium on a soybean leaflet (below). (Courtesy J. S. Mignucci)

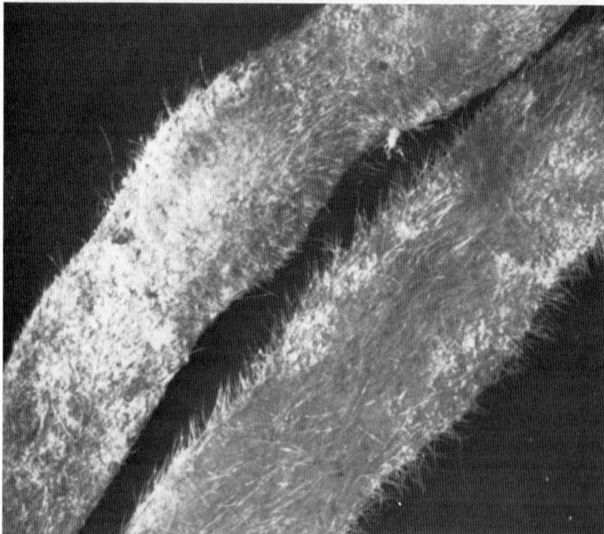

Fig. 24. Powdery mildew: leaf spots and lesions (above) and pod lesions (below) covered with powdery masses of conidiophores and conidia of *Microsphaera diffusa*. (Courtesy D. V. Phillips [above] and J. S. Mignucci [below])

Fig. 26. Conidiophores and conidia typical of *Microsphaera diffusa*, the cause of powdery mildew of soybeans, in a scanning electron micrograph ($\times$140). (Courtesy N. M. Kokalis)

Epidemiology

The disease usually develops in cooler than normal seasons, when photosynthesis and transpiration are reduced. Under controlled conditions, temperatures of 18 and 24°C favored disease development, and the disease was arrested at 30°C.

Resistance governed by a dominant gene, *Rmd*, has been incorporated into a large number of commercial cultivars.

Control

1. Use resistant cultivars where available.
2. Some foliar fungicides may control powdery mildew, but they may not increase yield, and they may require too many applications to be economical.

Selected References

Buzzell, R. I., and Haas, J. H. 1978. Inheritance of adult resistance to powdery mildew in soybeans. Can. J. Genet. Cytol. 20:151-153.

Dunleavy, J. M. 1980. Yield losses in soybeans induced by powdery mildew. Plant Dis. 64:291-292.

McLaughlin, M. R., Mignucci, J. S., and Milbrath, G. M. 1977. *Microsphaera diffusa*, the perfect stage of the soybean powdery mildew pathogen. Phytopathology 67:726-729.

Mignucci, J. S., and Boyer, J. S. 1979. Inhibition of photosynthesis and transpiration in soybean infected by *Microsphaera diffusa*. Phytopathology 69:227-230.

Mignucci, J. S., and Chamberlain, D. W. 1978. Interactions of *Microsphaera diffusa* with soybeans and other legumes. Phytopathology 68:169-173.

Mignucci, J. S., and Lim, S. M. 1980. Powdery mildew development on soybeans with adult-plant resistance. Phytopathology 70:919-921.

Mignucci, J. S., Lim, S. M., and Hepperly, P. R. 1977. Effects of temperature on reactions of soybean seedlings to powdery mildew. Plant Dis. Rep. 61:122-124.

Phillips, D. V. 1984. Stability of *Microsphaera diffusa* and the effect of powdery mildew on yield of soybean. Plant Dis. 68:953-956.

(Prepared by Julia S. Mignucci)

Red Leaf Blotch

Red leaf blotch of soybeans has been reported in central to southern Africa, from Cameroon east to Ethiopia and south to Zimbabwe. Estimated yield losses of 7 to 37% in approximately 25% of the soybean-producing area of Zambia have been reported, along with 10 to 50% yield losses in certain provinces of Zimbabwe.

Symptoms

The red leaf blotch fungus causes lesions on foliage, petioles, pods, and stems. Lesions are often associated with primary leaf veins. They may initially occur on unifoliolate leaves and are dark red to brown, circular to angular, and 1 to 3 mm in diameter. When trifoliolate leaves fully expand, the lesions are dark red on the upper surface of the leaves and reddish brown on the lower surface. They enlarge and coalesce to form large necrotic blotches, up to 2 cm in diameter (Fig. 27). Heavily diseased plants defoliate and senesce earlier than normal. Within older lesions, sclerotia develop primarily on the lower surface of leaves, and pycnidia develop primarily on the upper surface.

Causal Organism

Red leaf blotch is caused by *Dactuliochaeta glycines* (R. B. Stewart) Hartman & Sinclair (syn. *Pyrenochaeta glycines* R. B. Stewart); its sclerotial state is *Dactuliophora glycines* Leakey. Within the lesions on leaves of soybeans and the wild soybean *Neonotonia wightii* (Arnott) Lackey (a perennial legume), pycnidia are 87 to 298 μm in diameter, with aseptate to multiseptate setae clustered mostly around the ostiole.

Conidiogenous cells are monophialidic and ampulliform and are formed from the inner cells of the pycnidial wall. Conidia are ellipsoidal, one-celled, and 4-8 × 1-3 μm. Sclerotia are 96 to 357 μm in diameter, mostly spherical, and dark brown to black, with setae 5 to 36 μm long.

Disease Cycle

The disease cycle has been characterized. Sclerotia from the soil and those that land on leaf surfaces give rise to pycnidia, either on their surfaces or possibly from mycelium. The pycnidia form conidia, which can infect, and sclerotia on leaves can form mycelium that penetrates directly. Within older lesions, both pycnidia and sclerotia develop.

Control

No commercial procedures are recommended for the control of red leaf blotch. However, control with fungicides and the use of resistance have been demonstrated experimentally.

Selected References

Datnoff, L. E., Naik, D. M., and Sinclair, J. B. 1987. Effect of red leaf blotch on soybean yields in Zambia. Plant Dis. 71:132-135.

Hartman, G. L., Datnoff, L. E., Levy, C., Sinclair, J. B., Cole, D. L., and Javaheri, R. 1987. Red leaf blotch of soybeans. Plant Dis. 71:113-118.

Hartman, G. L., and Sinclair, J. B. 1988. *Dactuliochaeta* a new genus for the fungus causing red leaf blotch of soybeans. Mycologia 80:696-706.

(Prepared by G. L. Hartman)

Rhizoctonia Aerial Blight

Rhizoctonia aerial blight of soybeans has been reported in subtropical and tropical areas. First reported in the Philippines in 1918, it has since been reported in the southern part of the People's Republic of China, India, Malaysia, Mexico, Puerto Rico, Taiwan, and the Gulf Coast area of the United States. Yield losses of 35% have been attributed to the disease.

Besides soybeans, the causal fungus attacks common bean (*Phaseolus vulgaris* L.), lima bean (*P. lunatus* L.), clovers (*Trifolium* spp.), cowpeas (*Vigna* spp.), fescue (*Festuca* spp.), fig (*Ficus* spp.), lespedeza (*Lespedeza* spp.), rice (*Oryza sativa* L.), wild soybean (*Neonotonia wightii* (Arnott) Lackey), tung (*Aleurites* spp.), and many other genera.

Symptoms

Symptoms appear on leaves, stems, and pods, beginning on the lower or middle parts of the plant and moving up. Infected

Fig. 27. Red leaf blotch lesions, caused by *Dactuliochaeta glycines*, on the upper surface of soybean leaflets. (Reprinted, by permission, from Hartman et al, 1987)

leaves are water-soaked at first; they soon take on a greenish brown to reddish brown appearance and later become tan, brown, or black (Fig. 28). Lesions may be small spots, or the whole leaf may be blighted. Diseased tissues in old lesions that are no longer expanding generally fall out during dry weather, creating a ragged, shot-hole effect. Severely affected plants may be totally defoliated.

Infected leaves drop and adhere to pods and stems beneath, thus becoming sources of infection for pods and seeds. Brownish lesions form on petioles and stems. Lesions on pods may be small, brownish spots or may blight the entire pod, which often aborts. Seed infection is associated with pod infections.

Causal Organism

For details concerning the causal fungus, *Rhizoctonia solani* Kühn, see *Rhizoctonia* Diseases, under Fungal Diseases of Roots and Lower Stems. *Rhizoctonia* spp. are ecologically specific and differ morphologically. Isolates that cause foliage blight of soybeans may not necessarily cause root and stem rot. The aerial blight fungus is in anastomosis group 1 and produces both macrosclerotia and microsclerotia.

Disease Cycle and Epidemiology

The disease occurs in geographic areas characterized by prolonged periods of high humidity and warm weather. Symptoms may appear at any stage of plant development. Sclerotia, which survive in the soil for extended periods, serve as primary inoculum.

Control

1. Seed treatment may limit early-season disease development.

2. Apply a foliar fungicide when the disease first appears.
3. Use resistant cultivars where available.

Selected References

Berggren, G. T., McGawley, E. C., Pace, M. E., Gershey, J. S., and Joye, G. F. 1984. Strategies for controlling soybean aerial blight in Louisiana. (Abstr.) Phytopathology 74:872.

Hepperly, P. R., Mignucci, J. S., Sinclair, J. B., Smith, R. S., and Judy, W. H. 1982. Rhizoctonia web blight of soybeans in Puerto Rico. Plant Dis. 66:256-257.

Horn, N. L., and Fontenot, M. F. 1980. Aerial blight of soybeans in Louisiana. South. Soybean Dis. Work. Proc. 7:117.

Joye, G. F. 1987. Management of Rhizoctonia aerial blight of soybean and biology of sclerotia of *Rhizoctonia solani*. Ph.D. dissertation, Louisiana State University, Baton Rouge. 91 pp.

Parmeter, J. R., Jr., ed. 1970. *Rhizoctonia solani*: Biology and Pathology. University of California Press, Berkeley. 255 pp.

(Prepared by G. T. Berggren, Jr., and J. P. Snow)

Rust

Soybean rust is widely distributed in the Eastern Hemisphere from Japan to Australia and westward to India (Fig. 29). It can be especially destructive in soybeans growing in subtropical and tropical areas of Asia. Yield losses as high as 40% have been reported in Japan, and losses of 10–50% have been reported in southern China, 10–40% in Thailand, and 23–90% in Taiwan. Nearly complete yield losses can occur in limited areas in most of these countries. The causal fungus has been reported on at least 64 leguminous hosts in the Eastern Hemisphere.

In the Western Hemisphere, soybean rust was reported first in an experimental planting in Puerto Rico in 1976 and later in Brazil and Colombia. However, the causal fungus was known to occur in many Latin American countries before 1976 (Fig. 30) and has been reported on at least 40 leguminous hosts. It has also been reported on soybeans in Tanzania and Zambia and on cowpea (*Vigna unguiculata* (L.) Walpers) in several other countries of tropical Africa.

Symptoms

The most common symptom is tan to dark brown or reddish brown lesions with one to many erumpent, globose, ostiolate uredia (Fig. 31 and Plate 6). The disease begins with small, water-soaked lesions, which gradually increase in size, turning from gray to tan or brown. They assume a polygonal shape restricted by leaf veins and may eventually reach a size of

Fig. 28. Symptoms of aerial blight, caused by *Rhizoctonia solani*, on a leaf petiole (above) and leaflets (below) of soybeans. (Courtesy C. R. Valles)

Fig. 29. Geographic distribution of soybean rust, caused by *Phakopsora pachyrhizi*, in the Eastern Hemisphere. (Courtesy J. B. Sinclair)

2–5 mm². Lesions can appear on petioles, pods, and stems but are most abundant on leaves, particularly on the abaxial surface. The number of uredia per lesion increases with increasing lesion age. Uredospores are exuded through the central pore in the uredium, forming clumps of spores above and around the uredium (Fig. 32).

Lesion coloration, gray-green to dark brown, is dependent on lesion age and the interaction between the soybean genotype and the race of the pathogen. When observed 14 days after inoculation, dark reddish brown lesions with zero, one, or two uredia and extensive necrosis (RB lesions) indicate a semi-compatible interaction, and tan lesions with two or more uredia and without extensive necrosis (TAN lesions) indicate a compatible interaction (Plate 7). Incompatible interactions are observed as small, grayish flecks. Lesions frequently are associated with leaf chlorosis, and high lesion densities result in premature defoliation and early maturity.

In the early stages of development (Plates 6 and 7), before the onset of sporulation, rust lesions may be confused with bacterial pustules (see Bacterial Pustule). However, the two diseases can be differentiated by the presence of multiple uredia in the rust lesion and by the irregular crack usually appearing in the host tissue of the bacterial pustule lesion.

Causal Organism

Soybean rust is caused by *Phakopsora pachyrhizi* Syd. & P. Syd. (syns. *P. meibomiae* (Arth.) Arth., *P. sojae* (P. Henn.) Sawada, *P. vignae* (Bres.) Arth., *Physopella concors* (Arth.) Arth., *Physopella meibomiae* Arth., *Uredo concors* Arth., *Uredo sojae* P. Henn., *Uredo vignae* Bres., *Uromyces sojae* (P. Henn.) Syd., and *Uromyces sojae* Miura non Syd.). In the Eastern Hemisphere, it was first identified on *Glycine soja* Sieb. & Zucc. as *Uredo sojae* by P. Hennings in 1903 and on *Pachyrhizus erosus* (L.) Urban as *P. pachyrhizi* by H. Sydow and P. Sydow in 1914. In the Western Hemisphere, it was first identified as *Uredo vignae* by G. Bresadola in 1891 and transferred to *P. vignae* by J. C. Arthur in 1917.

Spermagonia (pycnia) and aecia are unknown. Uredia are globose, subepidermal, erumpent, light cinnamon to reddish brown, and opened through a central pore. They form abundantly on the lower surface of leaves, where they range from 100 to 200 μm in diameter (Figs. 31 and 33 and Plates 6 and 7). They are fewer and smaller on the upper surface. Within lesions, new uredia develop continuously for several weeks. Paraphyses, united at the base, form a domelike covering over the sporophores. The paraphyses curve in, are hyaline to subhyaline and prominently capitate at the apex, have a narrow lumen, and measure about 7–15 μm toward the apex.

Uredospores are globose, subglobose, ovate, or ellipsoidal and are essentially hyaline to light yellow-brown (Fig. 32). The size of the spores is highly variable, in the range of 18–45 × 13–28 μm, depending on the host and environmental conditions. Germ pores are not obvious in the finely echinulate, hyaline wall (1–1.5 μm thick) of the uredospore. Uredospores can be exuded as a column through the central pore of the uredium.

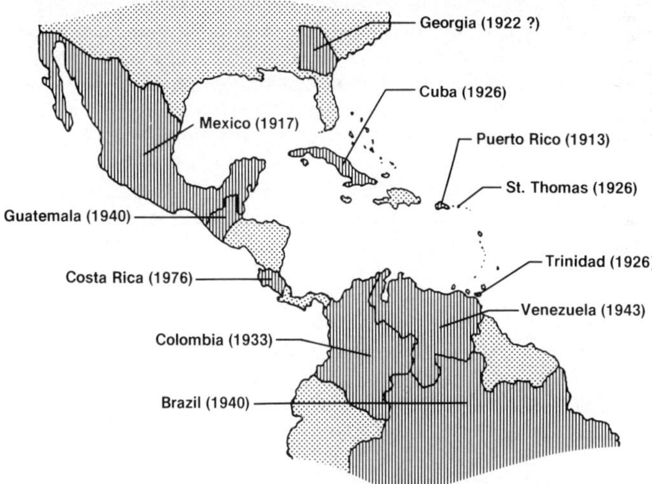

Fig. 30. Geographic distribution of *Phakopsora pachyrhizi*, the cause of soybean rust, in the Western Hemisphere. (Courtesy J. B. Sinclair)

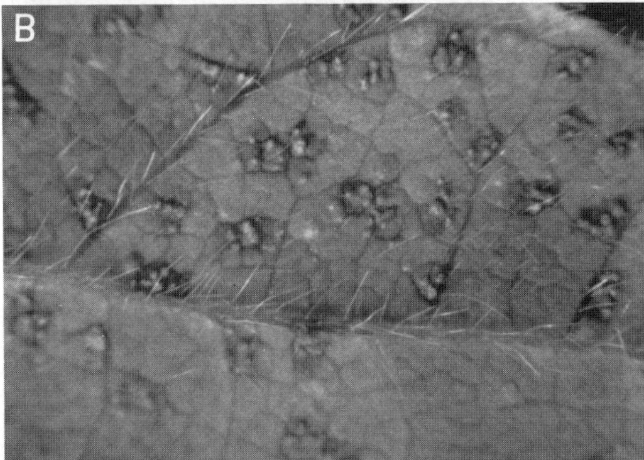

Fig. 31. **A,** Symptoms of soybean rust, caused by *Phakopsora pachyrhizi*, on the upper surface (left) and lower surface (right) of leaflets. **B,** Rust lesions and pustules on the lower surface of a leaflet. (Courtesy P. Sangawongse [A] and K. R. Bromfield [B])

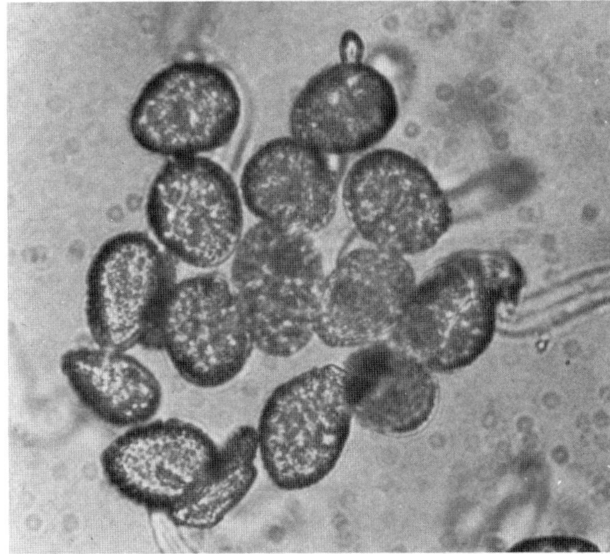

Fig. 32. Uredospores of *Phakopsora pachyrhizi*, the causal fungus of soybean rust, in a bright-field photomicrograph. (Courtesy P. Sangawongse)

Telia form subepidermally, mostly on the lower surface of leaves, among the uredia and at the edges of lesions (Fig. 33 and Plate 7). Orange-brown or light brown when young, they become dark brown to black with age. They are crustose, irregular to round, sparse to aggregated, and about 150–250 μm in diameter, with three to five irregular layers of teleutospores. Telia have been reported on 14 hosts and can be induced in many of these if the infected plant parts are kept at temperatures below 21°C for 2–4 weeks.

The teleutospores are yellow to brownish, one-celled, smooth-walled, and mostly clavate, oblong, or angulate. The spore wall is about 1 μm thick, except at the apex of outer spores, where it may be up to 5 μm thick and darker-colored. These spores vary in size, in the range of 13–35 × 5–15 μm. Germination and production of basidia and basidiospores have been reported. However, their role in the life cycle of the fungus and the epidemiology of the disease is unknown.

Disease Cycle and Epidemiology

The most intense rust epidemics have been observed in areas where the mean daily temperature is less than 28°C, with precipitation or long periods of leaf wetness occurring frequently throughout the growing season. Rust development is inhibited by dry conditions, excessive precipitation, or daily mean temperatures greater than 30°C or less than 15°C. The disease reduces photosynthetic activity by destruction of leaf tissue, premature defoliation, and reduction of the number of days to maturity. The cumulative effect of rust on yield is lower seed weight and fewer pods and seeds.

Free water on plant surfaces is necessary for uredospore germination and penetration. Uredospores germinate, usually by means of a single germ tube, over a temperature range of 8–28°C, with a broad optimum of 15–25°C. Upon germination, an appressorium roughly the size of the parent spore develops on the end of the germ tube. Penetration occurs directly through the cuticle and underlying epidermal cell. After traversing the epidermal cell, the hyphae grow intercellularly. Entry through stomata has also been reported.

The optimal temperature for infection is 20–25°C. In this range, infection can occur within 6 hr after spore deposition; the success of infection is enhanced as the duration of leaf wetness increases. At 20°C, infected plants show chlorotic or tan-colored flecks about 5 days after inoculation. In 7 to 9 days, uredia are differentiated, and by 9 to 10 days, they begin liberating uredospores. A single uredium may produce uredospores for about 3 weeks. New uredia continue to develop within a lesion for about 7–8 weeks.

The host range of the pathogen includes at least 87 species in 40 genera, so far restricted to papilionaceous legumes. However, it is not known whether all of these hosts allow propagation of the fungus under natural conditions. The role that individual hosts play in soybean rust epidemics needs further investigation. Rust may persist throughout the year in these collateral hosts, which act as reservoirs of inoculum for soybeans. No evidence of seed transmission has been found.

Four dominant, independently inherited genes for resistance to *P. pachyrhizi*—Rpp_1, Rpp_2, Rpp_3, and Rpp_4—have been identified, in PI 200492, PI 230970, PI 462312 (Ankur), and PI 459025, respectively. These cultivars, as well as seven others, are suspected of containing genes for resistance. PI 239871A and PI 239871B (*G. soja*), PI 230971, PI 459024, Taita Kaohsiung No. 5, Tainung No. 4, and Wayne have been proposed as a set of differentials to identify races. By means of this set, nine races were identified at the Asian Vegetable Research and Development Center, in Taiwan. However, the predominant race was compatible with 10 of the 11 differentials, and all races were compatible with three or more of the differentials. Apparently, some rust races in the field already possess multiple virulence factors to known and suspected genes for resistance.

Genes for resistance also occur among the wild *Glycine* spp. from Australia. Efforts are being made to transfer this resistance to *G. max*.

Rate-reducing resistance has been demonstrated; however, it is difficult to evaluate because the rate of rust development is dependent on soybean development and maturity. Currently, evaluation for this type of resistance cannot be completed until after maturity; this limitation prevents the use of this method for screening large populations.

Evaluating soybeans on the basis of their ability to sustain lower rust-induced yield losses is a promising method of developing improved cultivars. This method has been used at the Asian Vegetable Research and Development Center to develop resistant cultivars that produce more stable yields across different environments.

Control

In addition to cultivar resistance, the application of fungicides can reduce rust damage. Mancozeb is widely used as a protectant spray. However, frequent applications are necessary for highly effective control, and the spray schedule has to be initiated before symptoms appear. Triadimefon also gives good control and can be applied less frequently than mancozeb.

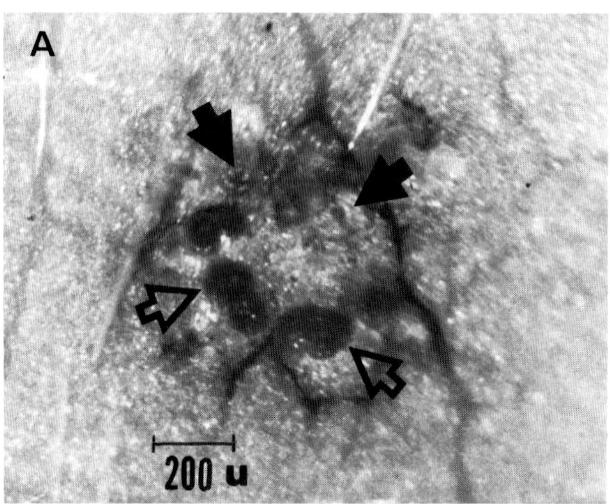

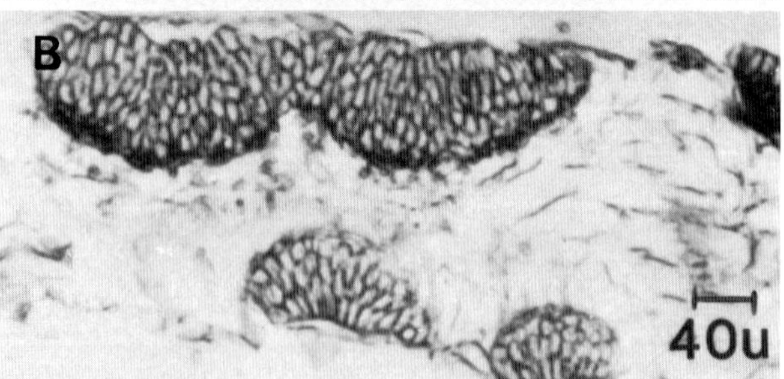

Fig. 33. Fruiting structures of *Phakopsora pachyrhizi*, the causal fungus of soybean rust, in bright-field photomicrographs. **A,** Telia (open arrows) and uredia (solid arrows) on a leaflet of PI 230970. **B,** Telia and teliospores formed on the upper and lower surfaces of a leaflet of PI 230971. (Courtesy C. C. Yeh)

Selected References

Bromfield, K. R. 1984. Soybean Rust. Monogr. 11. American Phytopathological Society, St. Paul, MN. 65 pp.

Ford, R. E., and Sinclair, J. B. 1977. Rust of Soybean; the Problem and Research Needs. INTSOY Ser. 12. College of Agriculture, University of Illinois at Urbana-Champaign. 110 pp.

Hartwig, E. E. 1986. Identification of a fourth major gene conferring resistance to soybean rust. Crop Sci. 26:1135-1136.

Kitani, K., and Inoue, Y. 1960. Studies on the soybean rust and its control measures. I. Studies on the soybean rust. (In Japanese; English summary) Shikoku Agric. Exp. Stn., Bull. 5:319-342.

Koch, E., and Hoppe, H. H. 1987. Germination of the teliospores of *Phakopsora pachyrhizi*. Soybean Rust Newsl. 8:3-4.

Saadaoui, E. M., and Tschanz, A. T., eds. 1987. Bibliography of

Soybean Rust, 1985–1986. Asian Vegetable Research and Development Center, Shanhua, Taiwan. 103 pp.

Tschanz, A. T., and Shanmugasundaram, S. 1985. Soybean rust. Pages 562-567 in: Proc. World Soybean Res. Conf. III. R. Shibles, ed. Westview Press, Boulder, CO.

Vakili, N. G., ed. 1978. Proc. Workshop Soybean Rust West. Hemisphere. U.S. Department of Agriculture, Agricultural Research Service, Mayaguez Institute of Tropical Agriculture, Puerto Rico. 81 pp.

Yeh, C. C., Sinclair, J. B., and Tschanz, A. T. 1982. *Phakopsora pachyrhizi*: Uredial development, uredospore production and factors affecting teliospore formation on soybeans. Aust. J. Agric. Res. 33:25-31.

(Prepared by A. T. Tschanz)

Scab

Scab, also called Sphaceloma scab, was discovered in Japan in 1947 and later named *Kokuto-byo* (scab). It has been reported only in Japan, where it is considered destructive. The causal fungus infects soybean leaves, stems, and pods.

Symptoms

Lesions on leaves are frequently visible on both surfaces and are commonly centered on veins. The lesions are circular to irregular, up to 4 mm in diameter, usually more or less raised above the leaf surface, and buff, often fading to a drab pale gray.

Stem lesions vary from minute, elliptical-elongate spots to large areas, up to 2 cm long, which develop as a result of the coalescence of smaller lesions. Stem lesions are buff, sometimes with reddish brown margins.

Young lesions on pods are red to brown (Fig. 34). At maturity, the lesions turn dark olive to black, with pale centers and reddish brown margins. Seeds fail to develop in severely infected pods.

Causal Organism

The causal fungus is *Sphaceloma glycines* Kurata & Kuribayashi. Acervuli are solitary or confluent; subcuticular and later erumpent; and pseudoparenchymatous at the stroma-like base. Conidiophores are hyaline at first, becoming dusky or dark with age, and are unbranched, blunted or pointed at the tips, continuous or one-septate, and closely compacted. They measure up to 26×4 μm and bear conidia on the tips. The conidia are ovoid or oblong-elliptical; hyaline, becoming dusky; biguttulate; and $4–13 \times 2–5$ μm (averaging 5×2 μm). The teleomorph of the fungus is not known.

Disease Cycle and Epidemiology

Direct infection occurs through the epidermis of leaf blades, petioles, and stems by means of an infection peg formed beneath an appressorium. After entering leaves, hyphae grow both among and within cells in the palisade and spongy mesophyll tissues. Chloroplasts and nuclei collapse, and cytoplasm coagulates and turns brown. In both young and old leaves of resistant cultivars, infection is confined to the epidermal cells. In mature leaves, abnormal cell division in the palisade tissue results in a cork layer three to five sclerenchymatous cells thick, which acts as a barrier against further extension of the developing mycelium.

Control

Soybean strains with resistance have been found, but resistant commercial cultivars have not been developed.

Selected References

Jenkins, A. E. 1951. Sphaceloma scab, a new disease of soybeans discovered by plant pathologists in Japan. Plant Dis. Rep. 35:110-111.

Kurata, H., and Kuribayashi, K. 1954. Soybean scab caused by *Sphaceloma glycines* sp. nov. (In Japanese) Ann. Phytopathol. Soc. Jpn. 18:119-121.

Tasugi, H., and Mogi, S. 1958. Resistance of soybean leaves to the scab, caused by *Sphaceloma glycines*. (In Japanese) Ann. Phytopathol. Soc. Jpn. 23:159-164.

Target Spot

Since 1945, when it was first reported in the United States, target spot has been found in most soybean-growing states. Yield losses of 18–32% have been recorded for very susceptible soybean lines grown in Mississippi during years when rainfall was above normal in August and September. The disease has also been reported in Brazil, Cambodia, Canada, the People's Republic of China, Hungary, Japan, Nicaragua, and the Soviet Union. Target spot is considered potentially serious, especially in late-maturing cultivars.

Symptoms

The disease affects leaves, stems, pods, seeds, hypocotyls, and roots. Leaf lesions are round to irregular and reddish brown and vary from specks to mature spots 10–15 mm or more in diameter (Fig. 35). Lesions are frequently surrounded by a dull green or yellowish green halo. Larger spots are often distinctly zonate; hence the common name, *target spot*. Narrow, elongated spots develop along the veins on the upper surface of

Fig. 34. Pod lesions caused by the soybean scab fungus, *Sphaceloma glycines*. (Courtesy J. B. Sinclair)

Fig. 35. Target spot lesions, caused by *Corynespora cassiicola*, on soybean leaflets. (Courtesy U.S. Department of Agriculture)

leaves of susceptible cultivars. Severely affected leaves drop prematurely.

Infected areas on petioles and stems are dark brown and range from specks to elongated, spindle-shaped lesions. Pod spots generally are circular and about 1 mm in diameter, with slightly depressed, purple-black centers and brown margins. During periods of extended rainfall or high humidity, spots may coalesce and cover the entire pod. In some cases, the causal fungus penetrates the pod wall and produces small, blackish brown lesions on the seeds.

Dark reddish brown, circular to oval lesions form on the hypocotyl, taproot, and large lateral roots, sometimes girdling the side roots. The lesions turn dark violet-brown when the fungus sporulates. They elongate and coalesce as the soybean plant ages, frequently girdling the taproot and stem. Entire roots may be discolored. Dark masses of conidiophores and conidia of the causal fungus form on the root surfaces as well as on other infected plant parts.

Causal Organism

Target spot is caused by *Corynespora cassiicola* (Berk. & Curt.) Wei (syns. *Cercospora melonis* Cooke, *Cercospora vignicola* Kawamura, *Helminthosporium vignae* Olive, and *H. vignicola* (Kawamura) Olive). At least two races of the fungus are known. The fungus that infects the hypocotyl, roots, and stem differs pathologically and morphologically from the fungus that infects the leaves, pods, and seeds and is probably a different species.

The erect, branched, brown conidiophores arise singly or in clusters from a swollen basal cell. They are one- to 20-septate and measure $44-135 \times 4-11 \mu m$, most commonly being three- to five-septate and $125-200 \times 8 \mu m$. Conidia are borne singly or in chains of two to six. They are typically smooth and brownish and vary greatly in shape, being cylindrical or broadest at the base and tapering slightly toward the tip. They are three- to 20-pseudoseptate, straight or slightly curved, and $39-520 \times 7-22 \mu m$ (averaging $135 \times 8 \mu m$). They have a central hilum at the basal end. Conidia germinate readily over a wide range of temperatures on potato-dextrose agar or Czapek agar, producing germ tubes at one or both ends.

Chlamydospores, formed in older cultures, are hyaline, terminal or intercalary, and oval, measuring $16-30 \times 14-20 \mu m$.

C. cassiicola grows and sporulates well on potato-dextrose agar, Czapek agar, and V-8 juice agar at pH 6.5 under continuous light. The optimal temperature range for growth of most isolates is $18-21°C$ (the maximum is $34-39°C$; the minimum is $5-7°C$). Some isolates grow best at $28°C$. The mycelium in culture is white and flocculent or appressed at first; later, it turns dark gray and forms an olivaceous black mat. The dark color eventually extends throughout the medium.

Disease Cycle and Epidemiology

C. cassiicola overseasons in diseased soybean stems, roots, and seeds and can survive in fallow soil for more than 2 years.

The fungus can colonize a wide range of plant residues in soil as well as cysts of *Heterodera glycines* Ichinohe.

Leaf infection takes place only when the relative humidity is 80% or higher or when free moisture is present on the leaves. Dry weather inhibits the fungus in both leaves and roots.

Stems and roots are first infected during the early seedling stage. Underground symptoms can be seen as early as 3 days after emergence. Soil temperatures of $15-18°C$ are optimal for infection and disease development. Seedlings growing at $15°C$ show large lesions on the primary roots, and growth of secondary roots is retarded. At $20°C$, symptoms are less severe, and seedlings produce nearly normal root systems. No symptoms develop at temperatures above $20°C$. The disease is less severe on late-planted soybeans; however, it differs considerably among cultivars.

The causal fungus is cosmopolitan and is especially abundant in the tropics, where it often causes distinct leaf spots on a wide range of host plants. Economically important plants that are infected include cassava (*Manihot* sp.); castor bean (*Ricinus communis* L.); cotton (*Gossypium hirsutum* L.); cowpea (*Vigna unguiculata* (L.) Walpers); cucumber (*Cucumis sativus* L.); Florida velvetbean (*Mucuna deeringianum* (Bort) Merr.); bean (*Phaseolus vulgaris* L.); lima bean (*P. lunatus* L.); mung bean (*V. radiata* (L.) R. Wilczek); guar (*Cyamopsis tetragonoloba* (L.) Taub.); blue lupine (*Lupinus hirsutus* L.); yellow lupine (*L. luteus* L.); okra (*Hibiscus esculentus* L.); papaw, or papaya (*Carica papaya* L.); pepper (*Capsicum frutescens* L.); rubber (*Hevea brasiliensis* Muell. Arg.); sesame (*Sesamum indicum* L.); sicklepod (*Cassia tora* L.); tomato (*Lycopersicon esculentum* Mill.); velvetleaf (*Abutilon theophrastii* Medic.), and watermelon (*Citrullus vulgaris* Schrad.).

Control

1. Most adapted, high-yielding soybean cultivars grown in the southeastern United States are tolerant.

2. Fungicides have been recommended in some states. Resistance to benomyl has been reported.

Selected References

Boosalis, M. G., and Hamilton, R. I. 1957. Root and stem rot of soybean caused by *Corynespora cassiicola* (Berk. & Curt.) Wei. Plant Dis. Rep. 41:696-698.

Carris, L. M., Glawe, D. A., and Gray, L. E. 1986. Isolation of the soybean pathogens *Corynespora cassiicola* and *Phialophora gregata* from cysts of *Heterodera glycines* in Illinois. Mycologia 78:503-506.

Horn, N. L., Lee, F. N., and Carver, R. B. 1975. Effects of fungicides and pathogens on yields of soybeans. Plant Dis. Rep. 59:724-728.

Seaman, W. L., Shoemaker, R. A., and Peterson, E. A. 1965. Pathogenicity of *Corynespora cassiicola* on soybean. Can. J. Bot. 43:1461-1469.

Spencer, J. A., and Walters, H. J. 1969. Variations in certain isolates of *Corynespora cassiicola*. Phytopathology 59:58-60.

(Prepared by J. P. Snow and G. T. Berggren, Jr.)

Fungal Diseases of Roots and Lower Stems

The etiology and diagnosis of root and lower stem diseases are complex, because soybean seeds, seedlings, and plants are susceptible to a wide range of seedborne and soilborne pathogens. Soilborne pathogens can cause stand losses even when the highest-quality planting seed is used with a fungicide seed treatment. Soilborne pathogens are omnipresent in most fields, particularly where soybeans have been grown before. However, these pathogens cause economic losses only when conditions favor their development. For example, stand losses of up to 100% have been observed when soybean seeds were sown in moist soils with temperatures of $35°C$ or higher, whereas soybeans sown in the same area under cooler, drier

conditions or under a mulch had minimal losses.

More than one pathogen is often involved in root and lower stem diseases. Careful laboratory analyses must be conducted to determine accurately the pathogen (or pathogens) involved in a disease.

Selected Reference

Shatla, M. N., and Sinclair, J. B. 1981. Soilborne pathogens and damping-off of soybeans. Pages 132-135 in: Soybean Seed Quality and Stand Establishment. J. B. Sinclair and J. A. Jackobs, eds. INTSOY Ser. 22. College of Agriculture, University of Illinois at Urbana-Champaign.

Brown Stem Rot

Brown stem rot (Phialophora stem rot), first reported in Illinois in 1944, is widespread in midwestern and some southeastern states of the United States and in Canada. The disease has also been reported in Egypt, Japan, Mexico, and Yugoslavia. It has also been reported on adzuki bean (*Vigna angularis* (Willd.) Ohwi & Ohashi) in Japan.

The extent of yield reductions depends on environmental conditions, the cultivar, and the fungal strain. Losses of 17 to 25% may be incurred because of the reduced number and size of seeds and the lodging of infected plants. Losses are greatest when cool weather during pod fill is followed by hot, dry weather. Growth stage R_5 is optimum for single-point assessment of stem and foliar symptoms.

Symptoms

Plants infected before pod set show no external symptoms. Infection becomes apparent about mid-season. A dark reddish brown discoloration of the vascular elements and pith, extending up from the roots or crown, is revealed in longitudinally split stems (Fig. 36). The browning at first extends 5–8 cm above the soil line and is prominent at the nodes. The more extensive the discoloration, the greater the yield reduction. As plants mature, the browning becomes continuous throughout the stems of susceptible cultivars. This symptom should not be considered diagnostic, because other pathogens can cause stem browning under the same conditions.

External symptoms may appear late in the growing season. The lower part of infected stems turns dull brown; this symptom is followed by sudden blighting and drying of leaves, which may drop prematurely. In dry weather, the leaves may suddenly wilt about 20–30 days before normal maturity. Tissues between the veins turn brown, and a narrow green border may outline the veins for a few days (Fig. 37).

When viewed from the edge of the field, a severely infected group of plants has a brownish cast, resembling frost damage, in contrast to the yellow-green of a normally maturing soybean field.

Causal Organism

Brown stem rot is caused by *Phialophora gregata* (Allington & Chamberlain) W. Gams (syn. *Cephalosporium gregatum* Allington & Chamberlain). On soybean stem extract agar, conidiophores of *P. gregata* are hyaline, aseptate or septate, straight or club-shaped, and usually simple. They can be borne anywhere on the hyphae, usually appear in groups, and are 4–15 μm long, occasionally reaching 25 μm. Conidiophores borne on hyphal tips are not well differentiated. Hyphae (1.2–4.7 μm in diameter) are hyaline, septate, and branched (Fig. 38).

Conidia are produced in succession at the tips of the conidiophores and aggregate in a slime droplet, forming irregular, noncatenulate heads (Fig. 38). Newly formed conidial heads disintegrate instantly upon contact with water, unless the slime has dried and thus holds them in place. Aseptate, uniseptate, and multiseptate conidia may be seen in culture. One-celled conidia are ovoid to elliptical and hyaline. Those produced in soybean stems measure 3.9–4.3 × 6.8–9.4 μm; those produced on soybean stem extract agar are smaller, measuring 1.7–3.4 × 3.4–7.6 μm. Occasionally, conidia elongate within 24–36 hr of production, and a definite transverse septum forms in the middle of each. This process continues, forming conidia several times longer than the original ones and containing several septa. One end of a septate conidium may constrict to give rise to a budlike projection, which then separates into a daughter conidium similar to the one-celled type. An elongate conidium may produce several buds, but daughter conidia do not produce buds. Conidia germinate in distilled water within 24 hr; they usually form a single germ tube at one end, although bipolar germination may occur.

The growth medium greatly influences the cultural characteristics, growth rate, and sporulation of *P. gregata*. The fungus grows slowly and sporulates poorly in about 2 months on oatmeal, potato-dextrose, and potato-dextrose-raisin agars. Sporulation occurs within 2 to 3 weeks on bean pod, soybean stem, and water agars. Light is not essential for sporulation.

The optimal temperature for growth of the fungus on soybean stem agar is 22–24°C (the maximum is 30°C; the minimum is 8°C). The optimal temperature for sporulation is 15–20°C; no sporulation occurs at 28°C. The optimal temperature for conidial germination is 21–15°C (the maximum is 30°C; the minimum is 15°C). Sporulation is greatest at pH 5–6.

Isolates of *P. gregata* vary widely in cultural characteristics, growth rate in culture, sporulation, virulence, and to some

Fig. 37. Symptoms of brown stem rot, caused by *Phialophora gregata*, on soybean leaflets. (Courtesy L. E. Gray)

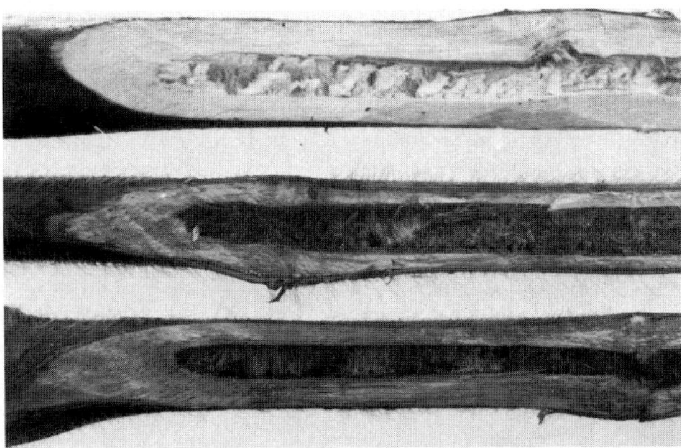

Fig. 36. Longitudinally split stem of an uninfected soybean plant (top) and stems of plants infected with *Phialophora gregata*, the brown stem rot fungus, causing discoloration of the vascular elements and pith. (Courtesy U.S. Department of Agriculture)

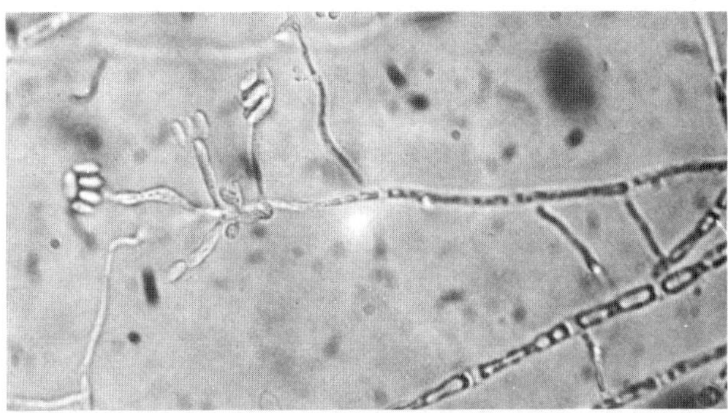

Fig. 38. Hyphae and conidia of *Phialophora gregata*, the cause of brown stem rot of soybeans, on 4% soybean stem decoction agar. (Courtesy L. E. Gray)

extent the type of symptoms produced in diseased plants. Highly virulent isolates can cause premature death of plants, which is not caused by less virulent isolates.

Disease Cycle and Epidemiology

P. gregata survives in soil and soybean debris and sporulates profusely on soybean debris buried in soil to a depth of 30 cm. The fungus sporulates on all types of soybean debris except pods and can do so several times on the same piece of straw, unless other microorganisms invade the straw. However, the source of primary inoculum is not known. Intensive soybean cultivation and the shift from long to short rotations with maize and other nonhost crops have contributed to increased distribution and severity of brown stem rot in the United States.

Mycelium of *P. gregata* usually enters plants through main and lateral roots and slowly grows upward in the xylem vessels. The spread of conidia from the xylem stream to other plant parts has been postulated. Mycelium within xylem cells varies from a few strands to a solid mass occupying the entire lumen.

Disease development is greatest between 15 and 27°C. As temperatures approach 27°C, internal browning lessens, and little or no disease develops at 32°C. Cool weather leads to more browning.

The progress of the disease depends on the physiological rather than the chronological age of the plant. The disease develops faster in older than in younger plants, regardless of temperature. Plants appear more susceptible when approaching the reproductive stage.

The fungus produces the wilt-inducing toxins gregatin A, C, and D.

Resistance has been reported in PIs 437833, 84946-2, 86150, 88820, and 90138. Resistance is governed by a single dominant gene, *Rbs*. Field resistance has been incorporated in the commercial cultivars BSR101, BSR201, BSR301, BSR302, and Chamberlain.

Control

1. Rotate soybeans with maize or another nonhost crop for three years.
2. Use resistant cultivars in areas where brown stem rot is a severe problem.

Selected References

Allington, W. B., and Chamberlain, D. W. 1948. Brown stem rot of soybean. Phytopathology 38:793-802.

Gams, W. 1971. *Cephalosporium*-artige Schimmelpilze (Hyphomycetes). Gustav Fischer Verlag, Stuttgart. 262 pp.

Gray, L. E. 1974. Role of temperature, plant age, and fungus isolate in the development of brown stem rot in soybeans. Phytopathology 64:94-96.

Hanson, P. M., Nickell, C. D., Gray, L. E., and Sebastian, S. A. 1988. Identification of two dominant genes conditioning brown stem rot resistance in soybean. Crop Sci. 28:41-43.

Kobayashi, K., Kondo, N., Ui, T., Tachibana, H., and Aota, T. 1983. Difference in pathogenicity of *Phialophora gregata* isolates from adzuki bean in Japan and from soybean in the United States. Plant Dis. 67:387-388.

Mengistu, A., and Grau, C. R. 1987. Seasonal progress of brown stem rot and its impact on soybean productivity. Phytopathology 77:1521-1529.

Schneider, R. W., Sinclair, J. B., and Gray, L. E. 1972. Etiology of *Cephalosporium gregatum* in soybean. Phytopathology 62:345-349.

Tachibana, H., and Card, L. C. 1979. Field evaluation of soybeans resistant to brown stem rot. Plant Dis. Rep. 63:1042-1045.

Taylor, S. L., Peterson, R. E., and Gray, L. E. 1985. Isolation of gregatin A from *Phialophora gregata* by preparative high-pressure liquid chromatography. Appl. Environ. Microbiol. 50:1328-1329.

(Prepared by L. E. Gray)

Charcoal Rot

Symptoms of charcoal rot (dry-weather wilt, or summer wilt) appear in hot, dry weather or when unfavorable environmental conditions stress the plant. The disease may appear in irrigated soybeans, along with accelerated maturity, when water is withheld after flowering. The pathogen attacks the plant throughout the season, often causing progressive debilitation of the host.

The causal fungus is widely distributed in soils, and the disease occurs worldwide. In the tropics, where the pathogen causes blight of emerging seedlings, plant losses up to 77% have been reported. In temperate or tropical areas, external and internal signs of the fungus usually appear anytime after growth stage R_1, especially in plants of low vigor or in a weakened condition. When severe, the disease reduces yields and seed quality.

Symptoms

Infected seedlings may show a reddish brown discoloration at the emerging portion of the hypocotyl. Twin-stem abnormality may develop. If infection occurs through the roots, discoloration is evident at the soil line and above. The discolored area turns dark brown to black, and infected seedlings may die, particularly under hot, dry conditions. If wet, cool weather persists, infected seedlings survive but carry a latent infection through the reproductive stages. Latent infection can be detected by treatment with paraquat. Symptoms may reappear later with hot, dry weather.

In older plants, the charcoal rot phase appears after midseason. Infected plants at first produce leaves smaller than normal and show a subtle loss of vigor. In a more advanced stage, leaves turn yellow and wilt but remain attached (Fig. 39A). Occasionally, superficial stem lesions extend upward from the soil line. After flowering, a light gray or silvery discoloration of the epidermal and subepidermal tissues develops in the taproot and the lower part of the stem. Microsclerotia form in the vascular elements and may block water flow (Fig. 39B). Frequently microsclerotia 50–75 μm in diameter are produced in the pithy area of the stem. They may be so numerous as to give a grayish black color to the tissues beneath the epidermis (Fig. 39C), resembling a sprinkling of finely powdered charcoal. Microsclerotia on aboveground parts are first visible in stem nodes, as profuse, small, black, randomly distributed specks. The fungus causes a reddish brown discoloration in the vascular tissues of the taproot and progresses up into the stem, discoloring the vascular and pith tissues of the stem (Fig. 39D). Splitting open the taproot reveals black streaks in the woody portion of the crown.

Causal Organism

The causal fungus, *Macrophomina phaseolina* (Tassi) Goid. (syns. *M. phaseoli* (Maubl.) Ashby, *Rhizoctonia bataticola* (Taub.) Briton-Jones, *Sclerotium bataticola* Taub., and *Botryodiplodia phaseoli* (Maubl.) Thirum.), infects more than 500 hosts, including crops and weeds. It is highly variable, with isolates differing in microsclerotial size and the presence or absence of pycnidia. The pycnidial stage is not common on soybeans but is common on garden beans and jute beans. Some strains produce pycnidia on culture media (Fig. 40), but not those attacking soybeans; others can be induced to do so with special techniques.

The jet black microsclerotia appear smooth and round to oblong or irregular (Fig. 39C). Their size and shape vary within an isolate and on different substrates. They are uniformly reticulate and are formed from aggregates of hyphal cells joined by a melaninlike material. Individual microsclerotia are composed of 50 to 200 or more individual cells, united by a septal pore in each cell.

Pycnidia, initially immersed in host tissues, are erumpent at maturity. They are more or less globose; membranous or subcarbonaceous; dark to grayish, becoming black with age; and generally 100–200 μm in diameter. The small, truncate ostiole may be inconspicuous or have a definite opening (Fig. 40).

Cultural and morphological characteristics may differ among and within isolates and with the age of cultures. The fungus grows well on potato-dextrose agar and produces abundant microsclerotia, often 75–150 μm in diameter, depending on the nutritional level of the substrate. The optimal temperature for growth in culture ranges from 28 to 35°C. Two selective media have been developed for isolating *M. phaseolina* from soil.

M. phaseolina produces colonies in culture that range from white to brown or gray and become darker with age. Aerial mycelium, with completely or partially appressed growth, may or may not be produced. Some isolates form concentric growth rings. Hyphal branches generally arise at right angles to parent hyphae, but branching at an acute angle is common. Branches may have a constriction at the point of union, which caused earlier workers to incorrectly classify this fungus as *Rhizoctonia. M. phaseolina* and *B. theobromae* Pat. have been confused in culture.

Disease Cycle and Epidemiology

Microsclerotia may survive in soil or embedded in host residue in dry soils for long periods. In wet soils, they cannot

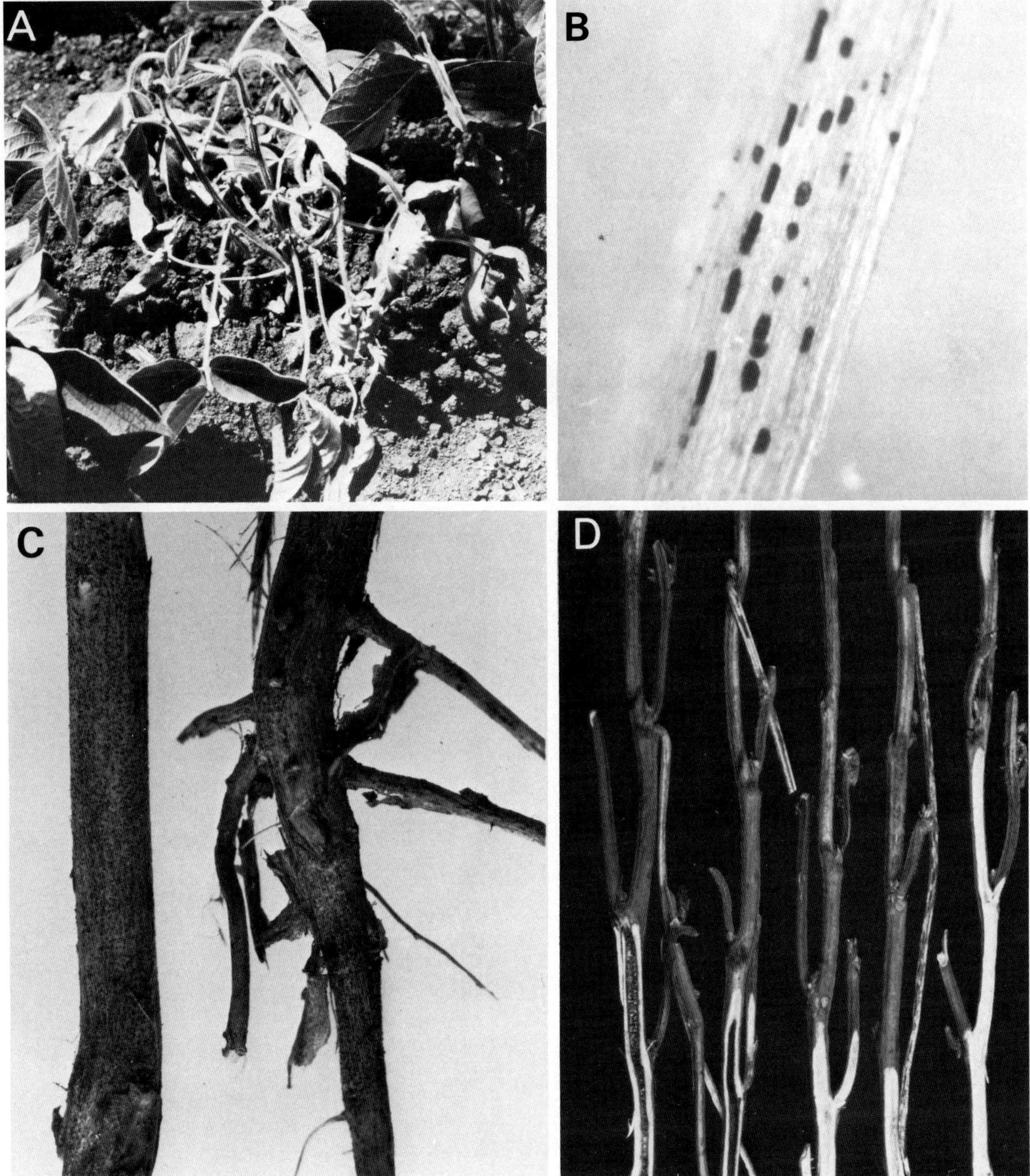

Fig. 39. Charcoal rot of soybeans. **A,** Wilting of a young plant. **B,** Microsclerotia in the vascular elements of a young plant. **C,** Small, black microsclerotia exposed after removal of the epidermis of infected plants. **D,** Black streaks in the woody tissues and microsclerotia in the pith of split stems of infected plants. (Courtesy M. B. Ilyas [A and B], U.S. Department of Agriculture [C], and H. J. Walters [D])

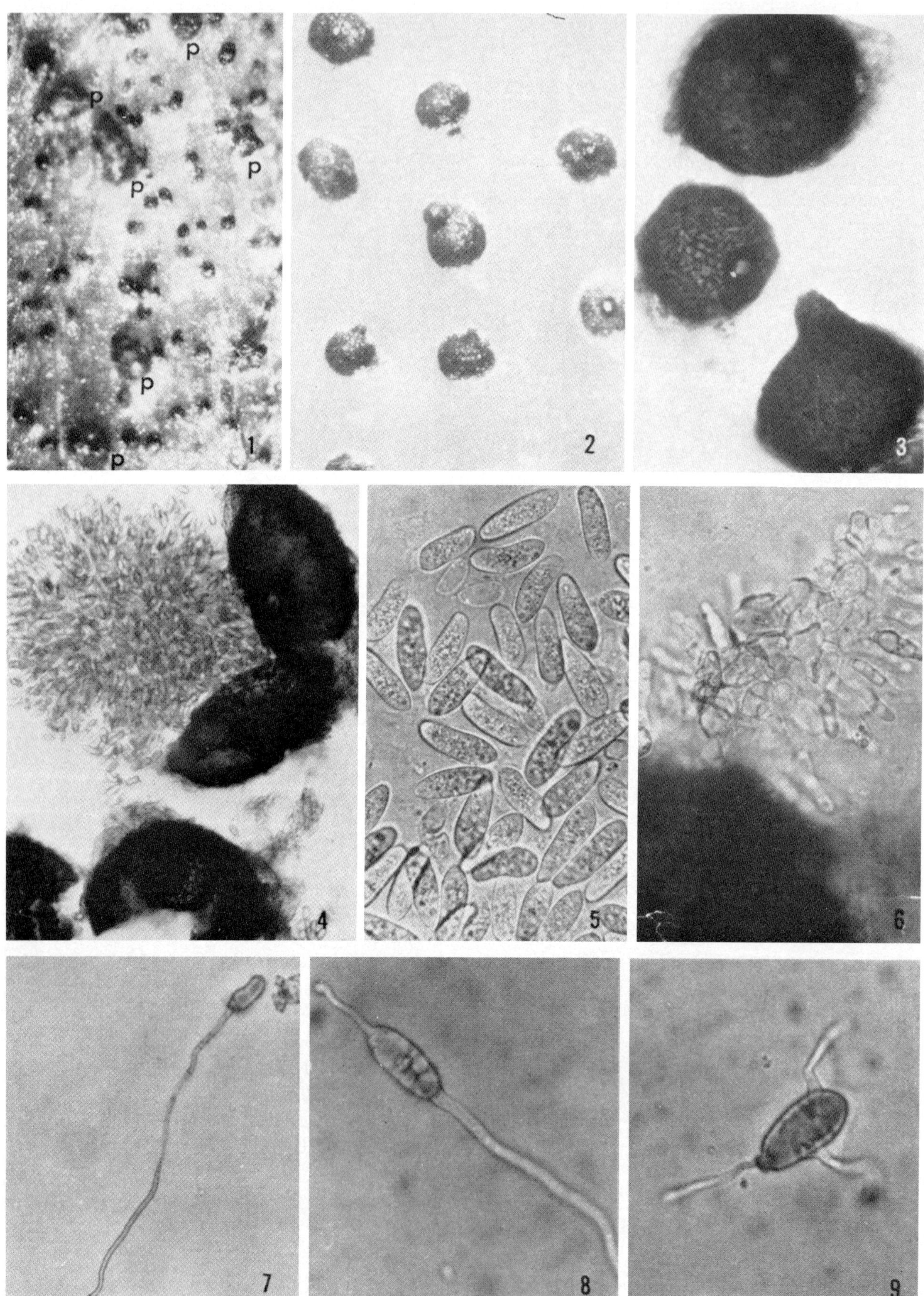

Fig. 40. Pycnidia and pycnidiospores of *Macrophomina phaseolina*, the causal fungus of charcoal rot. **1,** Pycnidia (p) on a stem segment of a dry bean plant (*Phaseolus* sp.). **2 and 3,** Pycnidia, with ostioles visible. **4,** Pycnidium releasing pycnidiospores. **5,** Pycnidiospores. **6,** Conidiophores in a pycnidium. **7–9,** Pycnidiospores germinating with a single germ tube (7) two germ tubes (8), and three germ tubes (9). (Reprinted, by permission, from Watanabe, 1972)

survive more than 7 to 8 weeks, and mycelium no more than 7 days. *M. phaseolina* competes well when soil nutrient levels are low and temperatures are above 30°C. Growth in soil is initiated by nutrients stored in microsclerotia and continues when soil nutrient levels are insufficient for fungal competitors. Soil populations increase when soybeans are grown continuously in the same field, and the disease becomes more severe in successive crops.

A large percentage of seeds may carry the pathogen in the seed coat. Infected seeds either do not germinate or produce seedlings that may die soon after emergence. (See *Macrophomina phaseolina*, under Seed Pathology.)

The disease is not evident at low temperatures; the pathogen begins to grow and symptoms appear between 28 and 35°C. Pathogen growth can occur early in the season, often infecting 80–100% of seedlings 2 to 3 weeks after planting. The rate of infection increases with higher soil temperatures. Low soil moisture further enhances disease severity.

Microsclerotia germinate on the root surface and produce numerous germ tubes. Penetration generally occurs from appressoria formed over anticlinal walls of epidermal cells or through natural openings. The fungal hyphae grow first intercellularly and then intracellularly through the xylem and form microsclerotia that plug the vessels (Fig. 39B). Microsclerotia form in green or juvenile tissues, usually as a result of moribundity and release of nutrients. *M. phaseolina* probably causes disease via the mechanical plugging of xylem vessels by microsclerotia, toxin production, enzymatic action, and mechanical pressure during the penetration of the middle lamellae.

Control

1. In severely infested fields, rotate with comparatively poor hosts, such as cereals or cotton, for one or two years; with maize or grain sorghum, rotation must be for three years.

2. Avoid excessive seeding rates. Crowding produces weakened seedlings, which are more vulnerable to fungal attack.

3. Fertilize soybeans to encourage vigorous growth.

4. Irrigate, where possible, to keep soil moisture high, or flood fields for 3 to 4 weeks before planting.

Selected References

Dhingra, O. D., and Sinclair, J. B. 1977. An Annotated Bibliography of *Macrophomina phaseolina* 1905–1975. Universidad Federal Vicosa, Vicosa, Brazil. 244 pp.

Dhingra, O. D., and Sinclair, J. B. 1978. Biology and Pathology of *Macrophomina phaseolina*. Universidad Federal Vicosa, Vicosa, Brazil. 166 pp.

Meyer, W. A., Sinclair, J. B., and Khare, M. N. 1973. Biology of *Macrophomina phaseoli* in soil studied with selective media. Phytopathology 63:613-620.

Mihail, J. D., and Alcorn, S. M. 1987. *Macrophomina phaseolina*: Spatial patterns in a cultivated soil and sampling strategies. Phytopathology 77:1126-1131.

Pearson, C. A. S., Leslie, J. F., and Schwenk, F. W. 1986. Variable chlorate resistance in *Macrophomina phaseolina* from corn, soybean, and soil. Phytopathology 76:646-649.

Pearson, C. A. S., Schwenk, F. W., Crowe, F. J., and Kelley, K. 1984. Colonization of soybean roots by *Macrophomina phaseolina*. Plant Dis. 68:1086-1088.

Short, G. E., Wyllie, T. D., and Bristow, P. R. 1980. Survival of *Macrophomina phaseolina* in soil and in residue of soybean. Phytopathology 70:13-17.

Watanabe, T. 1972. Pycnidium formation by fifty different isolates of *Macrophomina phaseoli* originated from soil or kidney bean seed. Ann. Phytopathol. Soc. Jpn. 38:106-110.

Wyllie, T. D., and Brown, M. F. 1970. Ultrastructural formation of sclerotia of *Macrophomina phaseolina*. Phytopathology 60:524-528.

(Prepared by T. D. Wyllie)

Fusarium Blight or Wilt, Root Rot, and Pod and Collar Rot

Fusarium-induced diseases of soybeans have been attributed to different species, form species, and physiologic races of *Fusarium*. These diseases occur in most soybean-growing areas of the world. Fusarium blight or wilt of soybeans was first reported in 1917, and Fusarium root rot in 1961. Economic losses of up to 59% resulting from blight or wilt, up to 64% from root rot, and up to 50% from reduced pod formation have been reported. Yields have reportedly been reduced by as much as 59%. Fusarium pod and collar rot was reported in India in 1972. Data on economic losses from this disease are not available; however, in a related study 40% of infected seeds planted failed to germinate.

Symptoms

Fusarium blight or wilt. At least three form species of *F. oxysporum* Schlecht. ex Fr. emend. Snyd. & Hans. can cause blight or wilt. Symptoms appear about mid-season in hot weather (at temperatures above 28°C), particularly on plants growing in sandy soils. The disease has not been reported in seedlings. The most characteristic symptom is browning or blackening of the vascular system in roots and stems, which is evident when stems are split open. Leaves of affected plants may become chlorotic, wither, and eventually drop. Flaccid leaves and wilting of stem tips are most common on young plants. Pods of infected plants are often poorly developed, but root rot is minor.

Fusarium root rot. *F. oxysporum* causes root rot, which usually develops on seedlings and young plants in cool weather (14°C). Older plants are generally less susceptible than younger ones. When the disease is severe, seedling emergence is slow and poor, and affected seedlings are stunted and weak. Infection is generally confined to the roots and lower stem (Fig. 41). Cotyledons of diseased seedlings are chlorotic and later become necrotic. The lower part of the taproot and lateral root system may be destroyed (Fig. 42). The pathogen is usually confined to the cortex, but vascular elements are invaded in advanced stages of Fusarium root rot. When soil moisture is low, infected seedlings or plants may wilt, and in some instances, plants in an entire field may be wilted.

Adventitious roots commonly develop from the upper (older) portion of the taproot, producing a shallow fibrous root system. Infected plants beyond the seedling stage seldom die, but their seeds are generally small and shriveled. Although the pathogen is normally confined to the roots, nearly mature pods may be

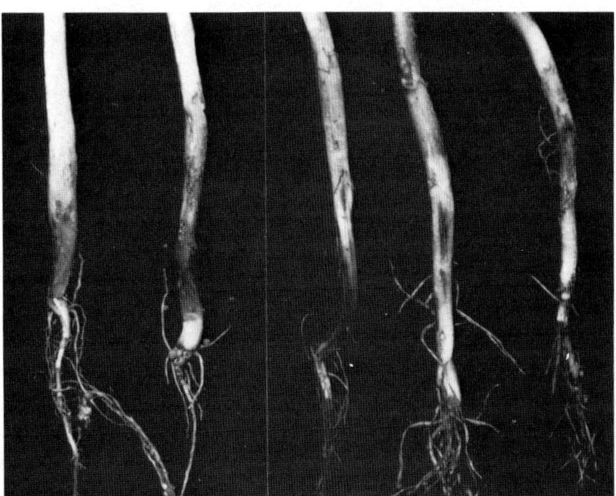

Fig. 41. Root and stem lesions caused by *Fusarium* spp. on soybean seedlings. (Courtesy J. M. Dunleavy)

invaded under prolonged humid, wet conditions. Pod infection may result in seed transmission of the pathogen.

Pod and collar rot. F. semitectum Berk. & Rav. causes pod and collar rot. Depressed, water-soaked, cream-colored, serrated lesions form on cotyledons and hypocotyls of emerging or slightly older seedlings. As the seedlings mature, these areas turn dark brown to black and eventually coalesce to form large lesions. Invaded hypocotyls are thin and soft; infected roots are underdeveloped. In advanced stages of pod and collar rot, fungal growth is evident on the infected tissues.

Pods may dry prematurely; the drying begins at the pod tip and progresses toward the base. Pods eventually turn dark brown or black. Severely infected pods produce no seeds. The seed coat and cotyledons of diseased seeds turn dark brown to black. This is particularly apparent in seeds in contact with infected pod tissue. Plants with decayed pods often show no other symptoms. *F. semitectum* is reported to be seedborne and to cause death of severely infected seeds in Brazil and Puerto Rico.

Causal Organisms

F. oxysporum, in the section Elegans, causes the blight or wilt and the root rot diseases of soybeans. Because the taxonomy of *Fusarium* is complex, other species, form species, and physiologic races have been reported as causal agents of these diseases. For example, *F. orthoceras* App. & Wollenw., now considered a synonym of *F. oxysporum*, was first reported as the cause of soybean root rot. In addition, *F. oxysporum* f. sp. *tracheiphilum* (E. F. Smith) Snyd. & Hans. race 1 (syn. *F. tracheiphilum* E. F. Smith) and *F. oxysporum* f. sp. *vasinfectum* (Atk.) Snyd. & Hans. have both been reported to cause blight or wilt of soybeans. Because of host specificity, isolates of *F. oxysporum* are differentiated on the basis of their pathogenicity on crop plants and classified into form species, which are morphologically and culturally indistinguishable. At least two physiologic races of *F. oxysporum* f. sp. *tracheiphilum*

Fig. 42. Fusarium root rot and lower stem decay of a soybean seedling (right), caused by *Fusarium* spp., and an uninfected seedling (left). (Courtesy U.S. Department of Agriculture)

have been reported; race 1 infects cowpea (*Vigna unguiculata* (L.) Walpers) and soybeans, whereas race 2 infects only cowpea. The cowpea cultivars Blackeye, Chinese Red, Climax, Grott, and Sumptuous are used to distinguish the two races. Race 2 of *F. oxysporum* f. sp. *vasinfectum*, the fungus causing wilt of cotton (*Gossypium* spp.), infects soybeans. *F. oxysporum* f. sp. *glycines* Armst. & Armst. is nonspecific.

F. oxysporum is highly variable in vitro, producing either aerial mycelium or a pionnotal form, which appears wet and slimy. In general, mycelium on potato-dextrose agar is white or tinged with purple, floccose, and septate; the undersurface is colorless, dark blue, or dark purple. Conidiophores are branched or unbranched monophiliades, usually short and dolioform. Sporodochia are cream, tan, or orange. Chlamydospores are intercalary or terminal and formed singly or in pairs. Microconidia are single-celled, oval to kidney-shaped, and produced in false heads. Macroconidia are hyaline, three- to five-septate, and sickle-shaped, with a foot-shaped basal cell and an attenuated apical cell.

F. semitectum, in the section Arthrosporiella, causes primarily pod and collar rot of soybeans. The fungus is variable in culture, producing either white mycelium or a pionnotal type. In general, aerial mycelium on potato-dextrose agar is tan to brown. Sporodochia, if present, are orange. The agar undersurface varies from peach to tan to brown. Conidiophores are unbranched or branched monophiliades and polyphialides. Chlamydospores are present and are intercalary or terminal. Microconidia are scarce. Macroconidia borne in aerial mycelium are spindle-shaped, with a papilla at the basal cell; those borne in sporodochia are slightly curved, with a foot-shaped basal cell.

F. equiseti (Corda) Sacc., *F. graminearum* Schwabe (teleomorph *Gibberella zeae* (Schw.) Petch), *F. moniliforme* (Sheld.) Snyd. & Hans. (teleomorph *G. fujikuroi* (Sawada) Ito), and *F. solani* (App. & Wollenw.) Snyd. & Hans. have also been reported to be pathogenic to soybeans.

The following have been isolated from soybeans in the Soviet Union: *F. culmorum* (W. G. Sm.) Sacc., *F. gibbosum* App. & Wollenw. emend. Bilai, *F. gibbosum* var. *acuminatum* (Ell. & Ev.) Bilai, *F. gibbosum* var. *bullatum* (Sherb.) Bilai, *F. heterosporum* Nees, *F. oxysporum* var. *orthoceras* (App. & Wollenw.) Bilai, *F. solani* var. *coeruleum* (Sacc.) C. Booth, *F. sporotrichiella* Bilai, and *F. sporotrichiella* var. *poae* (Pk.) Wollenw. emend. Bilai.

Disease Cycle and Epidemiology

F. oxysporum inhabits the soil by colonizing various plant residues and overseasons as chlamydospores or mycelium. Consequently, primary inoculum comes from soil. Seedborne infection by *F. oxysporum*, *F. rigidiusculum* (Brick) Snyd. & Hans., *F. semitectum*, *F. moniliforme* var. *subglutinans* Wollenw. & Reinking, and *F. tricinctum* (Corda) Sacc. has been reported and occasionally reduces seed germination. (See also *Fusarium* spp., under Seed Pathology.)

F. oxysporum f. sp. *tracheiphilum* penetrates soybean plants through wounds. In early stages of infection, it is confined to xylem vessels near the pith. Later, most xylem vessels become filled with mycelium, and parenchyma tissues also are invaded. *F. oxysporum* is most destructive when the soil is saturated with water at temperatures of 14–23° C; infection is greatest at the lower temperatures in this range. Penetration occurs directly through epidermal cells and indirectly through lenticels, wounds resulting from secondary root development, or stomates on the hypocotyl. Initial intercellular penetration is followed by intercellular growth of hyphae in the cortex. The stele is generally not affected; however, in advanced stages the xylem may be invaded.

The soybean cyst nematode (*Heterodera glycines* Ichinohe), a root-knot nematode (*Meloidogyne incognita* (Kofoid & White) Chitwood), and a sting nematode (*Belonolaimus longicaudatus* Rau) predispose seedlings and young soybean plants to infection by *F. oxysporum* (see Nematode Diseases), as do dinitroaniline herbicides. Seedlings growing in soil infested with *H. glycines* and *F. oxysporum* frequently develop Fusarium wilt symptoms. Soybean plants are less predisposed by *M. incognita* than by *H. glycines*, because infection by *M. incognita* is limited mainly to the root tip region, and the larvae migrate intercellularly, resulting in only slight cellular destruction. Larvae of *H. glycines*, however, invade more mature root tissue, migrate intracellularly, and cause extensive wounding, thus making soybean plants more susceptible to attack by *Fusarium* spp.

F. oxysporum also interacts with *Rhizoctonia solani* Kühn in causing a root rot of soybeans (see *Rhizoctonia* Diseases).

F. oxysporum has been isolated from 16 dicotyledonous weeds, representing 10 plant families commonly found in soybean fields. Symptoms of Fusarium blight were not observed; however, all isolates were pathogenic on soybeans.

Control

1. Grow cultivars resistant to *Fusarium* and to soybean cyst and root-knot nematodes.
2. Plant high-quality seeds in warm, well-drained soil.
3. Delay cultivation until soil moisture is adequate.
4. In fields with a history of *Fusarium*, ridge soil around the bases of plants to promote the development of adventitious roots from the stem base.

Selected References

Agarwal, D. K., and Sarbhoy, A. K. 1978. Physiological studies on four species of *Fusarium* pathogenic to soybean. Indian Phytopathol. 31:24-31.

Berton, O., and Porto, M. D. M. 1982. Etiologia da mancha en reboleira da soja. Agron. Sulriograndense 18:85-101.

Datnoff, L. E., and Sinclair, J. B. 1988. The interaction of *Fusarium oxysporum* and *Rhizoctonia solani* in causing root rot of soybeans. Phytopathology 88:771-777.

Dhingra, O. D., Sediyama, C., Carraro, I. M., and Reis, M. S. 1978. Behaviour of four soybean cultivars to seed-infecting fungi in delayed harvest. Fitopatol. Bras. 3:277-282.

Helbig, J. B., and Carroll, R. B. 1984. Dicotyledonous weeds as a source of *Fusarium oxysporum* pathogenic on soybean. Plant Dis. 68:694-696.

Leath, S., and Carroll, R. B. 1982. Screening for resistance to *Fusarium oxysporum* in soybean. Plant Dis. 66:1140-1143.

Nelson, P. E., Toussoun, T. A., and Marasas, W. F. O. 1983. *Fusarium* Species: An Illustrated Manual for Identification. Pennsylvania State University Press, University Park. 193 pp.

Saharan, G. S., and Gupta, V. K. 1972. Pod and collar rot of soybean caused by *Fusarium semitectum*. Plant Dis. Rep. 56:693-694.

Schlub, R. L., Lockwood, J. L., and Komada, H. 1981. Colonization of soybean seeds and plant tissue by *Fusarium* species in soil. Phytopathology 71:693-697.

Sumner, D. R., and Minton, N. A. 1987. Interaction of Fusarium wilt and nematodes in Cobb soybean. Plant Dis. 71:20-23.

Todt, P., Arnold, W. E., and Carson, M. L. 1981. Relationship of Fusarium root rot of soybean (*Glycine max*) to four dinitroaniline herbicides. Proc. North Cent. Weed Control Conf. 36:28-31.

(Prepared by L. E. Datnoff)

Phytophthora Rot

Phytophthora root and stem rot was first observed in northeastern Indiana in 1948 and in northwestern Ohio in 1951. The first published report of the disease in the United States came in 1955. Since that time Phytophthora rot has been reported in Australia, Canada, Hungary, Italy, Japan, the Soviet Union, and elsewhere in the United States.

The disease may cause plant losses and yield reductions of 100% in very susceptible soybean cultivars. The severity of loss depends on cultivar susceptibility, rainfall, drainage, soil type, and tillage. Phytophthora rot is most severe in poorly drained clay soils where surface water collects. It may also appear in lighter soils and on well-drained ground if the soil is saturated for an extended period when plants are young.

The only reported hosts of the Phytophthora rot fungus other than soybeans are three *Lupinus* spp. native to the United States. The fungus has been demonstrated to cause damping-off of alfalfa (*Medicago sativa* L.) and sweet clover (*Melilotus alba* Desr.) and to be pathogenic to snap bean (*Phaseolus vulgaris* L.) and cranesbill (*Geranium carolinianum* L.) in greenhouse inoculation tests. Considerable confusion on the host range of the pathogen causing Phytophthora rot of soybeans exists in the literature, because isolates from soybeans, alfalfa, and other hosts were considered the same species for a number of years. Also, there is one report of *Phytophthora parasitica* Dast. causing soybean stem rot, and this pathogen has a very wide host range.

Symptoms

Phytophthora rot may be found in soybeans at any stage of development. Seed rot and preemergence damping-off can occur in flooded soils, often being misidentified as water damage. Postemergence damping-off and seedling root and stem rot cause wilting and death of plants and further reduce stands.

Symptoms on older seedlings depend on the relative susceptibility or tolerance of the cultivar. In "low-tolerant" cultivars at the primary leaf stage, affected stems may appear water-soaked, leaves may turn yellow and wilt, and affected seedlings may die. On the other hand, in "high-tolerant" cultivars, the damage may be restricted to roots, and seedlings appear only stunted.

Older plants of low-tolerant cultivars may be killed more gradually. Symptoms consist of yellowing between veins and along margins of lower leaves. Upper leaves become chlorotic, and then the plant wilts completely (Fig. 43A). Wilted leaves commonly remain attached to the plant. Affected plants usually occur in groups in a row rather than singly. Wilting symptoms are accompanied by destruction of the lateral roots and taproots. A dark brown discoloration often progresses up the stem, often as high as 10 nodes before the plant wilts (Plate 8). Internally, the cortex and vascular tissues are discolored.

In older plants of high-tolerant cultivars, symptoms are confined to root rot of the branch roots. The plants are not killed by the fungus but may be stunted and slightly chlorotic, with symptoms similar to those of mild nitrogen deficiency or severe flooding. Occasionally these symptoms are accompanied by long, narrow, sunken, brown lesions that progress up one side of the stem. These mild symptoms, referred to as hidden damage, may reduce yield by as much as 40%. Hidden damage is readily evident if plants subjected to disease control treatments, such as application of a soil fungicide, or cultivars with near-isogenic resistance are available for comparisons in the same field.

Foliar blight by *Phytophthora* has been reported after heavy rains. Progressive, light brown lesions with yellowish margins can colonize entire leaflets of young susceptible plants. Lesions are severely restricted on older leaves, a phenomenon referred to as age-related resistance.

Causal Organism

P. megasperma Drechs. f. sp. *glycinea* Kuan & Erwin (syns. *P. megasperma* var. *sojae* A. A. Hildebrand and *P. sojae* Kaufmann & Gerdemann) is the causal agent.

Sporangiophores are simple and indeterminate. Typically, terminal sporangia (conidia) are obpyriform ($42–65 \times 32–53 \, \mu$m) and nonpapillate (Fig. 43C and D). Sporangia germinate by extruding zoospores into a thin, delicate, membranous vesicle, which quickly expands and ruptures. Zoospores sometimes remain trapped within sporangia and germinate there. The resulting germ tubes penetrate the sporangium wall. Sporangia may also germinate directly, thus functioning as conidia. Empty sporangia commonly proliferate internally, forming new sporangia within the old. Hyphal swellings may be formed.

The optimal temperature for zoospore production is 20°C (the minimum is 5°C). The zoospores are ovoid, bluntly pointed at one or both ends, and flattened on the sides. They have two flagella, one directed anteriorly and the other, four to five times longer than the first, directed posteriorly. Just before encystment, at the end of the motility period, which may last up to several days, zoospore movement becomes sluggish and jerky. The cysts germinate directly by producing germ tubes, which frequently swell to form an appressorium when they contact a solid surface. After a short time, normal growth of mycelium is resumed from the appressorium. Cysts sometimes germinate by producing secondary zoospores, leaving the cyst membrane behind. Rarely, a miniature sporangium is formed at the tip of the germ tube. The optimal temperature for direct germination is 25°C; for indirect germination, 14°C. Cysts germinate more vigorously in nutrient solution than in distilled water.

Sexual organs (antheridia and oogonia) develop in abundance on cornmeal, lima bean, and potato-dextrose agars. The antheridia are paragynous and amphigynous. The oogonia

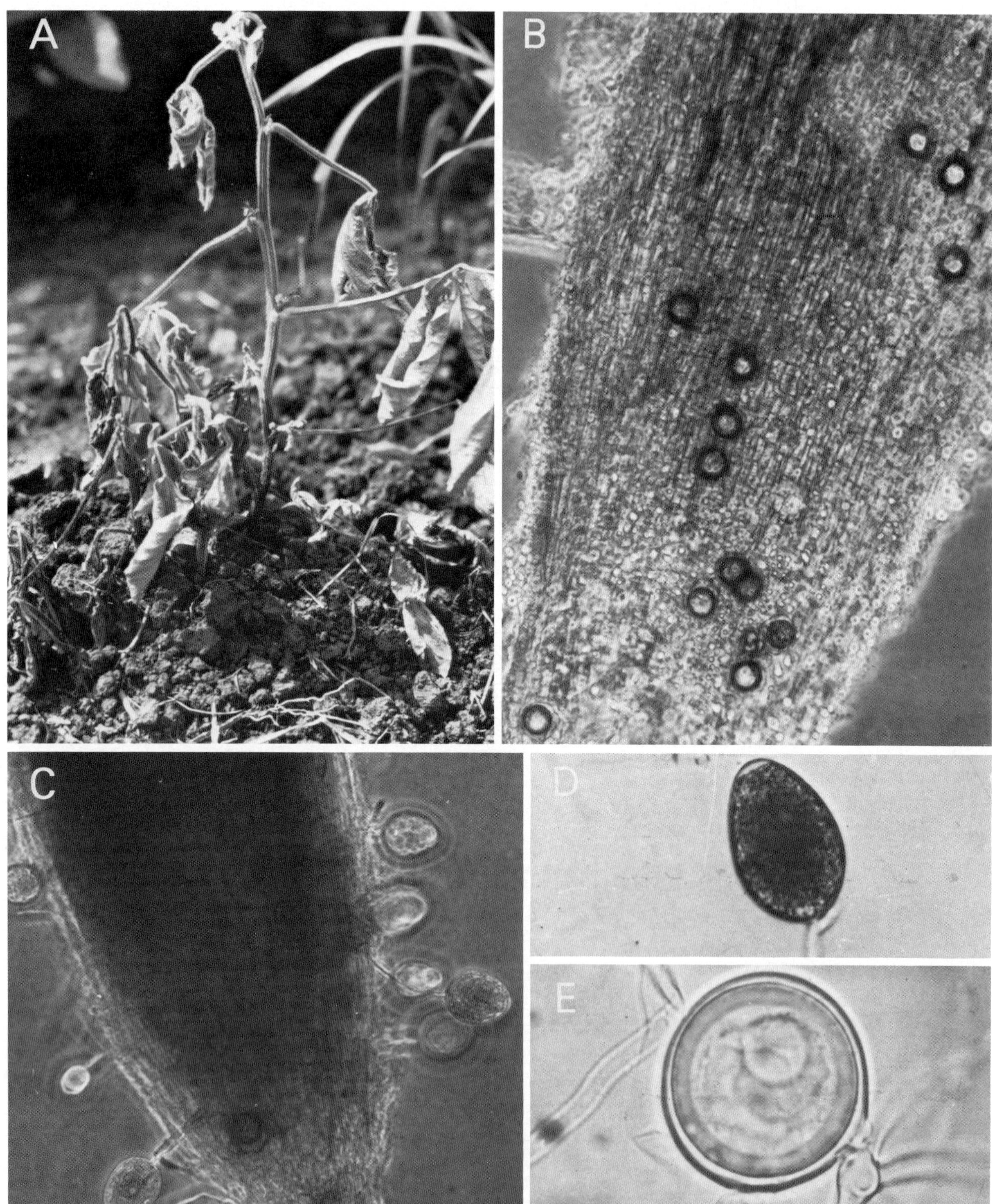

Fig. 43. Phytophthora rot of soybeans and structures of *Phytophthora megasperma* f. sp. *glycinea*, the causal fungus. **A,** Lower stem decay and death of a plant. **B,** Stained oospores in a root. **C,** Sporangia on a root tip. **D,** Dark sporangium (conidium). **E,** Light-colored oospore. (Courtesy W. A. Meyer [A], R. L. Slusher [B and C], and M. J. Kauffman and J. W. Gerdemann [D and E])

(29–58 μm in diameter) are thin-walled and spherical or subspherical. Oospores develop when an antheridium fertilizes an oogonium. Dormant oospores have thick, smooth inner and outer walls, fine-grained cytoplasm, a spherical refractive body in the center, a well-developed reserve granule, and a pair of pellucid bodies at the outer edge of the cytoplasm (Fig. 43B and E). After the dormant stage, but before germination, the pellucid bodies merge and fuse, and just before germination, they are not detectable. At the same time, the smooth inner wall erodes, and the central refractive body is absorbed. When an oospore germinates, the inner wall is completely absorbed, and a germ tube is produced, which develops into hyphae or a sporangium.

Antheridia, oogonia, and oospores develop abundantly on nutrient-rich media containing 1.5% or more agar. Abundant oospores are produced on agar containing lima bean extract (62.5 g of lima beans per liter), regardless of the agar concentration.

Oospores germinate about 30 days after formation. Germination occurs in distilled water but is improved by soil, low levels of nutrients, and root exudates. Light increases the percentage of oospore germination but inhibits the subsequent formation of sporangia. This inhibition of sporangial formation of is reversed by root exudates. Oospore germination is initiated within 2 days after separation from mycelium and exposure to appropriate conditions. The optimum temperature for formation and germination of oospores is 24° C.

Sporangia develop in dilute extracts of lima bean or other media. They can also be induced to form on solid media if washed repeatedly with water. Sporangia also form readily in young cultures grown on lima bean extract (50 g of frozen lima beans per liter) when washed repeatedly with Chen-Zentmyer salt solution. Oospore size and sporangial size and shape are affected by the medium and the age of the culture.

Mycelium is coenocytic when young, becoming septate with age; it branches mostly at right angles, with a slight constriction at the base of each branch. Hyphae are 3–9 μm wide and tend to curl. The optimum temperature for growth of most isolates is 25–28° C (the maximum is 32–35° C; the minimum is 5° C). Isolates of *P. megasperma* f. sp. *glycinea* vary widely in cultural characteristics, morphology, and virulence to soybean cultivars.

Two common selective media are used for isolating *P. megasperma* f. sp. *glycinea* from plants. One, P10VP, has a cornmeal agar base and contains 10 mg of pimaricin, 200 mg of vancomycin, and 100 mg of quintozene (pentachloronitrobenzene, PCNB) per liter. The second medium has a dilute V-8 juice agar base of 40 ml of V-8 juice neutralized by autoclaving with 0.6 g of calcium carbonate, filtered, diluted to 1,000 ml, and supplemented with 0.2 g of yeast extract, 1 g of sucrose, and 10 mg of cholesterol. Selective inhibitors for these media are 20 mg of 50% benomyl, 27 mg of 75% quintozene, 100 mg of neomycin sulfate, and 10 mg of chloramphenicol per liter. *P. megasperma* f. sp. *glycinea* cannot be readily isolated directly from soil. A semiquantitative baiting method has been described, utilizing soybean leaf disks floated on water flooded over soil for 1–2 hr and plated on selective media.

Disease Cycle and Epidemiology

Primary inoculum comes from crop residues in the soil, where the causal fungus survives as oospores for several years without soybeans. Many oospores are formed in the roots and stems of susceptible and tolerant soybean cultivars. How oospores germinate naturally in the soil has not been determined.

Phytophthora cannot be demonstrated in soil incubated for an extended period of time at 3° C or lower (conditions comparable to overwintering). However, after 1 week of incubation at 25° C, *Phytophthora* can be demonstrated in such soil by means of either seedlings or baits floated on flood water. From such observations, it is assumed that in the spring, oospores germinate whenever the temperature is suitable and form sporangia. Sporangia accumulate until the soil is flooded,

at which time zoospores are released. Sporangia also form on the surfaces of infected roots, providing secondary inoculum. Zoospores are produced in abundance in flooded and water-logged soils and are disseminated by soil water.

Zoospores near soybean roots are attracted toward the roots, where they encyst and germinate. Hyphae grow intercellularly in root tissues. Intracellular growth has been reported in hypocotyls, however. Globular and fingerlike haustoria penetrate host cells. There is no evidence of cell necrosis in advance of the pathogen. Colonization of host tissue in resistant interactions is similar to colonization in susceptible interactions for 24 hr. Then colonization ceases in resistant reactions.

Oogonia and oospores are formed in infected root and stem tissues of susceptible, tolerant, and resistant cultivars. Many more oospores are formed in susceptible and tolerant cultivars than in resistant ones. Although oospores form in resistant cultivars, little or no death or discoloration of roots occurs.

Leaf infection results in more severe foliar symptoms than does root infection. Infection occurs when soil particles containing the pathogen are deposited on leaves by wind or rainstorms. If the weather remains humid and cloudy, leaves become severely infected, and the fungus grows toward the petiole and stem.

Phytophthora root rot is most common in heavy, tightly compacted, fine-textured (clay) soils subject to flooding. Flooding rains within 1 week of planting are most favorable for disease development. High-tolerant plants may escape severe damage if they emerge and reach the first trifoliolate stage without conditions suitable for zoospore production. Low-tolerant cultivars can be damaged by mid- to late-season infection periods. *Phytophthora* damage can be reduced by tile drainage to remove excess moisture and reduce the duration of flooding. Reduced tillage, especially no-tillage, increases the damage. Monocropping of soybeans or application of a high level of fertilizer just before planting also may increase the damage.

The association and interaction of *P. megasperma* with other soil fungi may increase or decrease the severity of Phytophthora rot. Mycorrhizal fungi decrease infection of soybean roots. High populations of other root-rot fungi in the soil, such as *Fusarium* or *Pythium* spp. or *Rhizoctonia solani* Kühn, may increase the severity of Phytophthora rot. Infection of soybean plants by the northern root-knot nematode (*Meloidogyne hapla* Chitwood) also increases the severity of root rot.

Physiologic races and resistance. Two types of reaction to *Phytophthora* are found in soybeans. Race-specific resistance is controlled by at least 14 genes (*Rps* genes) at seven loci. Twenty-five races of *P. megasperma* f. sp. *glycinea* have been described, and many isolates have been found that do not fit any of the races. The races can be distinguished on eight soybean cultivars (Table 6).

Among susceptible cultivars, many quantitative differences in disease reaction can be found. The least susceptible reactions are referred to as field resistance, tolerance, or rate-reducing resistance. Tolerance levels can be evaluated by differential plant loss, vigor, and yield in infested soil in the field or greenhouse. Age-related resistance described in the hypocotyl and leaves may be similar.

Control

1. Use race-specific resistant cultivars. Cultivars need to be resistant to all races in the field.

2. Apply metalaxyl seed treatment for highly tolerant cultivars. Metalaxyl is an acylalanine fungicide that is specific for *Pythium* and *Phytophthora*. The seed treatment moves quickly into the cotyledons and top of the plant but not the roots. Therefore, it controls only the damping-off phase of infection by *Phytophthora*.

3. Combine optimum cultural conditions for integrated control of highly tolerant cultivars, good drainage, fall or spring plowing, and crop rotation. Metalaxyl seed treatment may still be necessary for controlling damping-off.

4. Apply metalaxyl in the seed furrow at planting time. Metalaxyl leaches into the root zone and is taken up by the roots before moving upward into the plant. *Phytophthora* can penetrate but cannot colonize roots containing metalaxyl at appropriate concentrations.

Selected References

Anderson, T. R. 1986. Plant losses and yield responses to monoculture of soybean cultivars susceptible, tolerant, and resistant to *Phytophthora megasperma* f. sp. *glycinea*. Plant Dis. 70:468-471.

Anderson, T. R., and Buzzell, R. I. 1982. Efficacy of metalaxyl in controlling Phytophthora root and stalk rot of soybean cultivars differing in field tolerance. Plant Dis. 66:1144-1145.

Beagle-Ristaino, J. E., and Rissler, J. F. 1983. Histopathology of susceptible and resistant soybean roots inoculated with zoospores of *Phytophthora megasperma* f. sp. *glycinea*. Phytopathology 73:590-595.

Bhattacharyya, M. K., and Ward, E. W. B. 1986. Expression of gene-specific and age-related resistance and the accumulation of glyceollin in soybean leaves infected with *Phytophthora megasperma* f. sp. *glycinea*. Biol. Mol. Plant Pathol. 29:105-113.

Canaday, C. H., and Schmitthenner, A. F. 1982. Isolating *Phytophthora megasperma* f. sp. *glycinea* from soil with a baiting method that minimizes *Pythium* contamination. Soil Biol. Biochem. 14:67-68.

Dirks, V. A., Anderson, R. T., and Bolton, E. F. 1980. Effect of fertilizer and drain location on incidence of Phytophthora root rot of soybeans. Can. J. Plant Pathol. 2:179-183.

Eye, L. L., Sneh, B., and Lockwood, J. L. 1978. Factors affecting zoospore production by *Phytophthora megasperma* var. *sojae*. Phytopathology 68:1766-1768.

Gray, L. E., and Pope, R. A. 1986. Influence of soil compaction on soybean stand, yield, and Phytophthora root rot incidence. Agron. J. 78:189-191.

Jimenez, B., and Lockwood, J. L. 1982. Germination of oospores of *Phytophthora megasperma* f. sp. *glycinea* in the presence of soil. Phytopathology 72:662-666.

Keeling, B. L. 1980. Research on Phytophthora root and stem rot: Isolation, testing procedures, and seven new physiological races. Pages 367-370 in: Proc. World Soybean Res. Conf. II. F. T. Corbin, ed. Westview Press, Boulder, CO.

Kotova, V. V. 1981. *Phytophthora megasperma* var. *sojae* Hild., an uncommon species in the USSR. Mikol. Fitopatol. 15:269-271.

Kuan, T.-L., and Erwin, D. C. 1980. Formae speciales differentiation of *Phytophthora megasperma* isolates from soybean and alfalfa. Phytopathology 70:333-338.

Layton, A. C., Athow, K. L., and Laviolette, F. A. 1986. New physiologic race of *Phytophthora megasperma* f. sp. *glycinea*. Plant Dis. 70:500-501.

Meyer, W. A., and Sinclair, J. B. 1972. Root reduction and stem lesion development on soybeans by *Phytophthora megasperma* var. *sojae*. Phytopathology 62:1414-1416.

Ploper, L. D., Athow, K. L., and Laviolette, F. A. 1985. A new allele at the *Rps3* locus for resistance to *Phytophthora megasperma* f. sp. *glycinea* in soybean. Phytopathology 75:690-694.

Schmitthenner, A. F. 1985. Problems and progress in control of Phytophthora root rot of soybean. Plant Dis. 69:362-368.

Schmitthenner, A. F., and Van Doren, D. M., Jr. 1985. Integrated control of root rot of soybean caused by *Phytophthora megasperma* f. sp. *glycinea*. Pages 263-266 in: Ecology and Management of Soilborne Plant Pathogens. C. A. Parker, A. D. Rovira, K. J. Moore, P. T. W. Wong, and J. F. Kollmorgen, eds. American Phytopathological Society, St. Paul, MN.

Tooley, P. W., and Grau, C. R. 1984. Field characterization of rate-reducing resistance to *Phytophthora megasperma* f. sp. *glycinea* in soybean. Phytopathology 74:1201-1208.

Tooley, P. W., Grau, C. R., and Stough, M. C. 1982. Races of *Phytophthora megasperma* f. sp. *glycinea* in Wisconsin. Plant Dis. 66:472-475.

Walker, A. K., and Schmitthenner, A. F. 1984. Heritability of tolerance to Phytophthora rot in soybean. Crop Sci. 24:490-491.

(Prepared by A. F. Schmitthenner)

Pod and Stem Blight and Phomopsis Seed Decay

Pod and stem blight was first observed in the United States in 1920 and has since been reported in Brazil, Canada, the People's Republic of China, Egypt, Guyana, India, Japan, Korea, Senegal, the Soviet Union, Taiwan, and Thailand. It is now endemic in most soybean-growing areas of the United States.

When soybeans mature in warm, wet weather and harvest is delayed, seed decay associated with pod and stem blight may lower seed viability. Diseased seeds may have a poor appearance and can lead to grade reduction and lower prices for producers. Infected seeds produce low-quality oil and flour. Seed quality aspects of Phomopsis seed infection associated with pod and stem blight are discussed in a separate section (see Phomopsis Seed Decay, under Seed Pathology).

The causal fungi may also occur on other crops and weeds. They colonize debris of pigeon pea (*Cajanus cajan* (L.) Millsp.); green bean (*Phaseolus vulgaris* L.); lima bean (*P. lunatus* L.); cowpea (*Vigna unguiculata* (L.) Walpers); garlic (*Allium sativum* L.); onion (*A. cepa* L.); lespedeza (*Lespedeza* spp.); lupines (*Lupinus* spp.); peanut, or groundnut (*Arachis hypogaea* L.); okra (*Hibiscus esculentus* L.); pepper (*Capsicum frutescens* L.); and tomato (*Lycopersicon esculentum* Mill.).

Symptoms

Stems, petioles, pods, seeds, and less frequently leaf blades may be infected (see Stem Canker; Phomopsis Seed Decay, under Seed Pathology). The pod and stem blight fungi first cause a latent infection throughout the plant. The pathogens can be readily isolated from apparently healthy tissue. Their presence can also be detected by placing detached green tissues in a moist chamber. Pycnidia develop in about a week. Surface sterilization and treatment of tissues with herbicides improve this detached tissue test. The pathogens can first be detected in the field as pycnidia on petioles of abscised leaves or on broken branches.

Dead stems may be covered with speck-sized pycnidia, usually arranged linearly (Fig. 44), or the pycnidia may be limited to small patches, generally near the nodes. In addition to the pod and stem blight pathogens, fungi associated with dead stems are species of *Alternaria*, *Colletotrichum* (Fig. 45), *Diaporthe*, *Phoma*, and *Septoria*.

Under field conditions, no definite leaf or stem lesions are produced. Pycnidia may be found in dead tissues, such as

Table 6. Responses of Differential Soybean Cultivars to Physiologic Races of *Phytophthora megasperma* f. sp. *glycinea*[a]

Differential Cultivar	*Rps* Gene	1	2	3	4	5	6	7	8	9	10	11	12	13	14	15	16	17	18	19	20	21	22	23	24	25
Harosoy	*Rpsh*	S	S	S	S	S	S	S	S	S	S	S	R	S	S	S	R	S	R	R	S	S	S	S	S	S
Harosoy 63	*Rps1*	R	R	S	S	S	S	S	S	S	R	R	R	R	S	R	R	R	R	R	S	S	S	S	R	S
Sanga	*Rps1-b*	R	S	R	R	R	R	R	R	R	S	S	S	R	R	R	S	R	R	S	S	R	R	S	S	S
Mack	*Rps1-c*	R	R	R	S	S	R	R	R	R	R	R	S	R	S	R	S	R	S	S	S	R	S	R	R	S
PI 103091	*Rps1-d*	R	R	R	R	R	S	R	S	R	R	R	S	R	R	R	R	S	R	S	R	R	R	R	R	R
Kingwa	*Rps1-k*	R	R	R	R	R	R	R	R	R	R	R	S	R	R	R	S	R	R	S	S	R	R	—	R	S
PI 171442	*Rps3*	R	R	R	R	R	S	S	R	R	S	R	S	R	R	S	R	S	R	S	S	S	S	R	S	R
Altona	*Rps6*	R	R	R	R	S	S	S	S	S	R	S	R	S	R	R	R	S	R	R	R	R	S	S	S	R

[a] R = resistant; S = susceptible.

Phytophthora-induced lesions, tissues damaged by hail injuries, or plants killed by other diseases. In wet seasons, latent infections produce pycnidia simultaneously over the entire plant after it matures. In dry weather, however, they are confined to limited areas on stems near the soil and are generally clustered close to nodes. Pycnidia are also scattered on dry, poorly developed pods (Figs. 44 and 45 and Plate 9). Not all infected pods produce pycnidia, but mature pods with pycnidia always contain infected seeds.

It has not been established that pod and stem blight stunts plants or reduces the number of seeds. The main losses are from the pod blight phase, resulting in moldy seeds, which may not be harvested, weigh less, or lead to grade reduction. Heavily infected seeds are badly cracked and shriveled and are frequently covered with white mycelium (Fig. 46 and Plate 10). Lightly infected seeds are often normal in size and appearance. Severely infected seeds often do not germinate.

Lesions on cotyledons of seedlings arising from infected seeds vary from almost colorless to bright red or brown and from pinpoint size to areas involving the whole cotyledon. The seed coat frequently remains attached after seedling emergence. Small reddish brown streaks, up to 1.5 cm long, occur on the hypocotyl at or below the soil line. Seedling blight may occur from seed infections. Buds may be deformed or completely decayed.

Causal Organisms

D. phaseolorum (Cke. & Ell.) Sacc. var. *sojae* (Lehman) Wehm. (anamorph *Phomopsis sojae* Lehman) (syn. *D. sojae* Lehman) and *Phomopsis longicolla* Hobbs are the two species involved in this disease. The relative prevalence of the two species varies with location and season. A third organism, *D. phaseolorum* var. *caulivora* Athow & Caldwell, is commonly associated with soybean seed decay but produces no pod and stem blight symptoms. It is the causal agent of stem canker (see Stem Canker).

Colonies of *D. phaseolorum* var. *sojae* on potato-dextrose agar are floccose and ropy, turning tan to brown as the culture ages. In reverse the colonies are tan to dark brown with black, pulvinate stromata. Conidiomata are pycnidial, black, stromatic, solitary or aggregated, and usually unilocular, with no beak or a beak less than 200 μm long, opening by an apical ostiole. Locules are uniostiolate to multiostiolate, lenticular, and up to 350 μm wide. Conidiophores are simple phialides; they are hyaline and up to $20 \times 1.5–2$ μm. Two types of conidia are produced (Fig. 47). Alpha conidia are hyaline, usually fusiform, guttulate, and $5.5–10.5 \times 1.5–3.5$ μm. Beta conidia, the more common type, are hyaline, filiform, and hamate.

Perithecia are produced on old agar cultures incubated in light or on overwintered soybean stems. Mature perithecia are nearly spherical and slightly flattened at the base ($148–282 \times 185–346$ μm). The perithecia have long, tapered beaks, $60–100 \times 60–150$ μm (Fig. 48). They are usually solitary (not clustered, unlike those of northern isolates of *D. phaseolorum* var. *caulivora*). Asci ($35–51 \times 3.3–10$ μm) are elongate and clavate and dissolve before ascospore liberation. The asci are

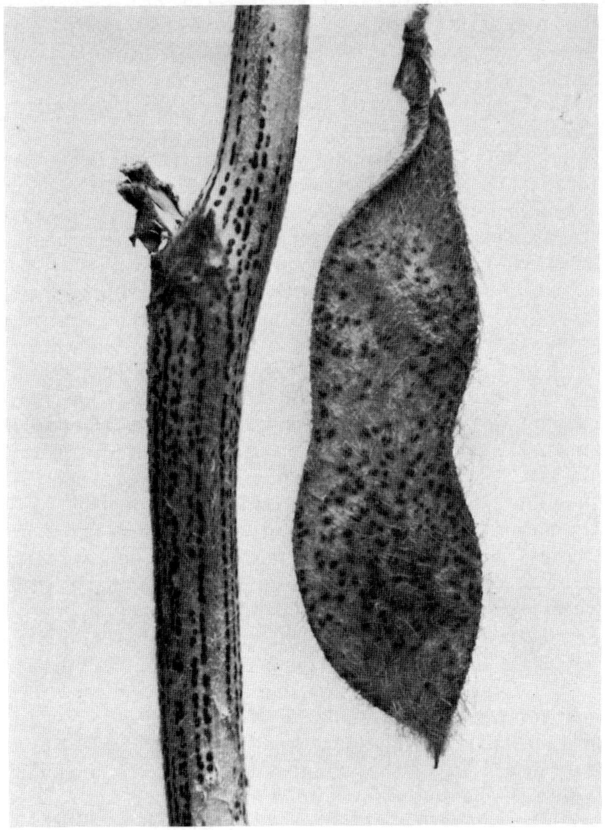

Fig. 44. Black pycnidia of *Diaporthe phaseolorum* var. *sojae* (*Phomopsis sojae*), the cause of pod and stem blight, on the lower stem and a pod of soybeans. (Courtesy U.S. Department of Agriculture)

Fig. 45. Symptoms produced on soybean pods and stem by *Diaporthe phaseolorum* var. *sojae* (*Phomopsis sojae*), the cause of pod and stem blight (left and center), and by the anthracnose fungus *Colletotrichum truncatum* (right). (Courtesy D. W. Chamberlain)

Fig. 46. Soybean seeds covered with growth of *Phomopsis longicolla*, the cause of Phomopsis seed decay. (Courtesy U.S. Department of Agriculture)

unitunicate, with a distinct apical ring. They are released in a viscous fluid that oozes out the ostiole tip. Little is known about how ascospores are dispersed; the ascus walls disintegrate rapidly, leaving free ascospores in the extrusion droplet. The ascospores are similar in shape to the alpha conidia (fusiform-elliptical) but are larger (9–13 × 2–6 μm) and bicellular. They are biguttulate in both cells; one germ tube may be produced at the apex of each cell.

P. longicolla, previously referred to as *Phomopsis* spp., is somewhat similar to *D. phaseolorum* var. *sojae* but has some important differences. Colonies on potato-dextrose agar are floccose, dense, and white with occasional greenish yellow areas. Cultures in reverse are colorless with large, black, spreading stromata. Conidiomata are pycnidial, black, stromatic, solitary or aggregated, and unilocular or multi-locular, with prominent necks more than 200 μm long, opening by an apical ostiole. Locules are uniostiolate or multiostiolate and globose, up to 500 μm wide. Conidiophores are hyaline, usually branched, septate, and 3.5–24 × 1–4 μm. Conidiogenous cells are hyaline, filiform, and phialidic. Alpha conidia are hyaline, ellipsoidal to fusiform, guttulate, and 5–9 × 1.5–3.5 μm. Beta conidia, which are rarely formed, are hyaline, filiform, and hamate.

Both *D. phaseolorum* var. *sojae* and *P. longicolla* grow well on acidified potato-dextrose agar (pH 5.0) at 28°C. The two species can be distinguished by mycelial characteristics, the size and shape of the stromata, pycnidial morphology, the shape of

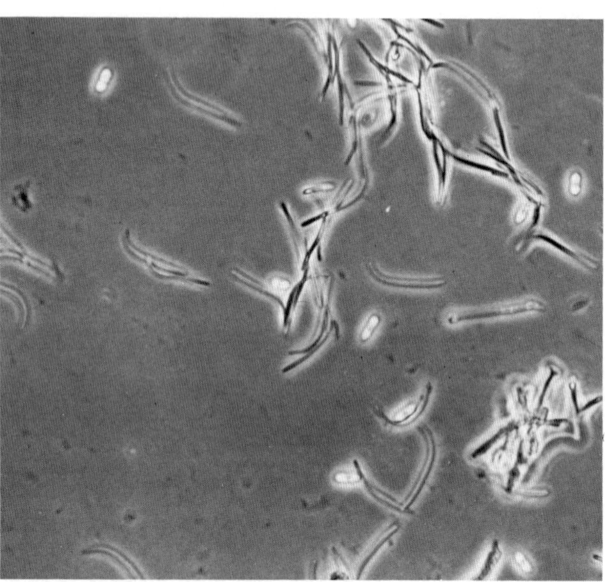

Fig. 47. Alpha conidia (one-celled and biguttulate) and beta conidia (elongate, filiform, and hamate) of *Phomopsis sojae*, the pod and stem blight fungus. (Courtesy F. D. Tenne)

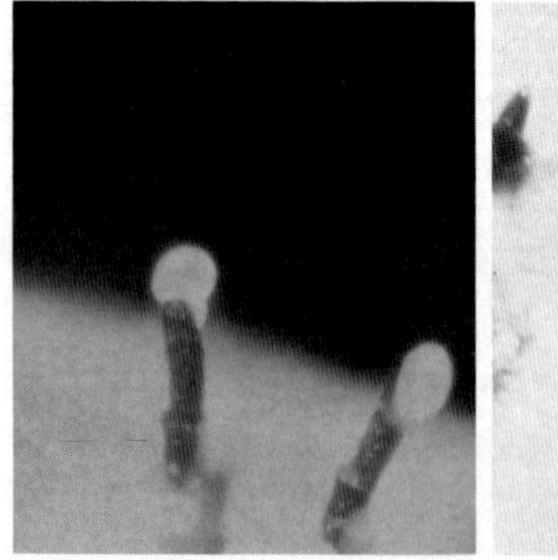
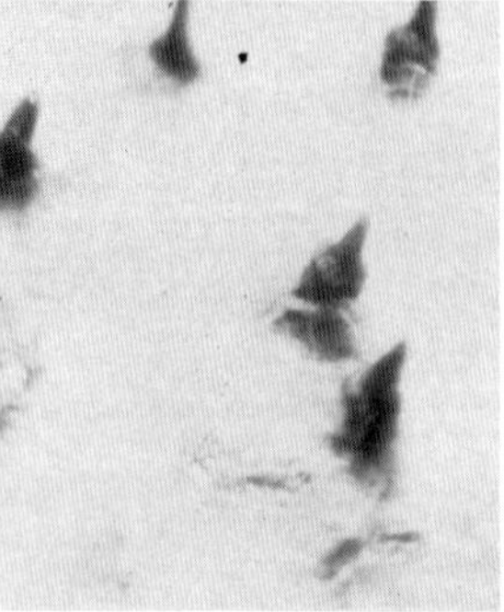

Fig. 48. Perithecia of *Diaporthe phaseolorum* var. *caulivora*, the cause of stem canker (left), and *D. phaseolorum* var. *sojae*, the cause of pod and stem blight (right). (Courtesy M. A. Crawford)

the pycnidial locule, conidiophore morphology, and the absence of perithecia in *P. longicolla.*

Disease Cycle and Epidemiology

The pod and stem blight fungi overseason as dormant mycelium in soybeans or other host debris and in infected seeds. Pycnidia of both *D. phaseolorum* var. *sojae* and *P. longicolla* are produced on overwintered debris and on petioles of the current year's abscised leaves. Perithecia of *D. phaseolorum* var. *sojae* are produced in early summer on overwintered stem debris.

The fungi colonize plant tissues within 2 cm of the point of infection until the plant begins to senesce; they then spread to tissues about 5 cm from the point of infection. Progressive spread in the plant is caused by infection from conidia dispersed by splashed water. Both alpha conidia and ascospores can be splashed onto plants and initiate infections, which are latent at first. Only infections that are initiated in pods can infect seeds and cause seed decay. Most seed infection occurs during or after the yellow pod stage (R_7). Prolonged wet periods during pod development and maturation and warm weather (temperatures above 20° C) favor the spread of the fungi from the pod to the seed. *P. longicolla* is more prevalent in seeds from pods at the bottom than at the top of the plant. More seed decay occurs in plants deficient in potassium, infected with a virus, or heavily attacked by insects. When harvest is delayed under wet conditions, seeds may be infected throughout the plant. Alternating periods of wet and dry weather favor pod deterioration and splitting and facilitate infection of mature seeds. Seed infection is greater in densely populated fields, because of excessive lodging of plants.

Seed decay is probably the most important phase of pod and stem blight. Diseased seeds are an important source of primary inoculum for long-range dissemination. The effects of *D. phaseolorum* var. *sojae* and *P. longicolla* on seed germination are discussed more fully under Seed Pathology. The stem phase of the disease is most important as a local source of inoculum.

Both *D. phaseolorum* var. *sojae* and *P. longicolla* are latent in immature host tissues. The mycelium is both intercellular and intracellular. Invasion is initially restricted to the thin-walled portion of the cortex. Later, the fungi penetrate the thick-walled xylem vessels. Pycnidia are arranged linearly, because of the tendency of the fungi to grow in the chlorenchyma lying between the resistant sclerenchyma tissues of the cortex. Mycelium invades the ovule and developing seeds through the funiculus and hilum. Within the seed, the fungi colonize all tissues of the seed coat and cotyledons and eventually the radicle and plumule. Symptomless infection of the seed coat may also occur.

The percentage of seeds infected with *Phomopsis* significantly decreases after a year or more of storage. Seeds free of *Phomopsis* can be obtained after extended storage.

Control

1. Rotate soybeans with corn, and plow down residues. The primary source of inoculum is infested straw from the previous crop.

2. Plant high-quality seeds relatively free of the pathogens, or use a fungicide seed dressing.

3. Plant late, or plant late cultivars. Late plantings and late cultivars generally have less seed infection than early plantings and early cultivars.

4. Use less susceptible cultivars. Good resistance has been found but not yet incorporated into good cultivars.

5. If possible, select a planting time that allows maturation during a dry period.

6. Use a foliar fungicide if the risk of seed infection is high, according to the point or pod infection prediction systems (see Soybean Disease Management Strategies). Fungicides applied to pods from mid-flowering (R_3–R_4) to the late pod stage (R_6–R_7) may lower the incidence of seed infection. Yield

responses from spraying are probably the result of control of foliar diseases.

7. Harvest soybeans promptly at maturity.
8. Maintain adequate potash to minimize moldy seed.

Selected References

Balducchi, A. J., and McGee, D. C. 1987. Environmental factors influencing infection of soybean seeds by *Phomopsis* and *Diaporthe* species during seed maturation. Plant Dis. 71:209-212.

Brown, E. A., and Minor, H. C. 1986. Characteristics of a soybean genotype resistant to Phomopsis seed decay. U.S. Dep. Agric., Soybean Genet. Newsl. 13:164-165.

Cerkauskas, R. F., and Sinclair, J. B. 1982. Effect of paraquat on soybean pathogens and tissues. Trans. Br. Mycol. Soc. 77:495-502.

Ellis, M. A., Ilyas, M. B., Tenne, F. D., Sinclair, J. B., and Palm, H. L. 1974. Effect of foliar application of benomyl on internally seed-borne fungi and pod and stem blight of soybean. Plant Dis. Rep. 58:760-763.

Garzonio, D. M., and McGee, D. C. 1983. Comparison of seeds and crop residues as sources of inoculum for pod and stem blight of soybeans. Plant Dis. 67:1374-1376.

Hepperly, P. R., Bowers, G. R., Jr., Sinclair, J. B., and Goodman, R. M. 1979. Predisposition to seed infection by *Phomopsis sojae* in soybean plants infected by soybean mosaic virus. Phytopathology 69:846-848.

Hepperly, P. R., and Sinclair, J. B. 1978. Quality losses in *Phomopsis*-infected soybean seeds. Phytopathology 68:1684-1687.

Hepperly, P. R., and Sinclair, J. B. 1980. Detached pods for studies of *Phomopsis sojae* pods and seed colonization. J. Agric. Univ. P.R. 64:330-337.

Hepperly, P. R., and Sinclair, J. B. 1980. Pod and stem blight disease rating and disease development at various heights on soybean plants. Crop Sci. 20:379-381.

Hobbs, T. W., Schmitthenner, A. F., and Kuter, G. A. 1985. A new *Phomopsis* species from soybean. Mycologia 77:535-544.

Horn, N. L., Carver, R. B., Lee, F. N., and Fort, T. M. 1979. The effect of timing of fungicide application on yield of soybeans. Plant Dis. Rep. 63:404-406.

Jeffers, D. L., Schmitthenner, A. F., and Kroetz, M. E. 1982. Potassium fertilization effects on Phomopsis seed infection, seed quality and yield of soybeans. Agron. J. 74:886-890.

Kmetz, K. T., Ellett, C. W., and Schmitthenner, A. F. 1979. Soybean seed decay: Sources of inoculum and nature of infection. Phytopathology 69:798-801.

Kmetz, K. T., Schmitthenner, A. F., and Ellett, C. W. 1978. Soybean seed decay: Prevalence of infection and symptom expression caused by *Phomopsis* sp., *Diaporthe phaseolorum* var. *sojae*, and *D. phaseolorum* var. *caulivora*. Phytopathology 68:836-840.

Kulik, M. M. 1984. Symptomless infection, persistence, production of pycnidia in host and non-host plants by *Phomopsis batatae*, *Phomopsis phaseoli* and *Phomopsis sojae*, and the taxonomic implications. Mycologia 76:274-291.

Kulik, M. M. 1984. Failure of *Phomopsis phaseoli* to produce mature pycnidia in senescent soybean stems at the end of the growing season. Mycologia 76:863-867.

Manandhar, J. B., Thapliyal, P. N., Kunwar, I. K., and Sinclair, J. B. 1987. Inhibition of *Colletotrichum truncatum* and *Phomopsis sojae* by their culture filtrates and maize or soybean stubble extracts. Biol. Cult. Tests 2:35.

McGee, D. C. 1986. Prediction of Phomopsis seed decay by measuring soybean pod infection. Plant Dis. 70:329-333.

Ross, J. L. 1986. Registration of eight soybean germplasm lines resistant to seed infection by *Phomopsis* sp. Crop Sci. 26:210-211.

Rupe, J. C., and Ferriss, R. S. 1986. Effects of pod moisture on soybean seed infection by *Phomopsis* sp. Phytopathology 76:273-277.

Shortt, B. J., Grybauskas, A. P., Tenne, F. D., and Sinclair, J. B. 1981. Epidemiology of Phomopsis seed decay of soybean in Illinois. Plant Dis. 65:62-64.

Sij, J. W., Turner, F. T., and Whitney, N. G. 1985. Suppression of anthracnose and Phomopsis seed rot on soybean with potassium fertilizer and benomyl. Agron. J. 77:639-642.

TeKrony, D. M., Egli, D. B., Balles, J., Tomes, L., and Stuckey, R. E. 1984. Effect of date of harvest maturity on soybean seed quality and *Phomopsis* sp. seed infection. Crop Sci. 74:189-193.

TeKrony, D. M., Egli, D. B., Stuckey, R. E., and Balles, J. 1983. Relationship between weather and soybean seed infection by *Phomopsis* sp. Phytopathology 73:914-918.

TeKrony, D. M., Stuckey, R. E., Egli, D. B., and Tomes, L. 1985.

Effectiveness of a point system for scheduling foliar fungicides in soybean seed fields. Plant Dis. 69:962-965.

(Prepared by A. F. Schmitthenner)

Stem Canker

Stem canker is capable of killing full-grown plants from mid-season to maturity. The disease has been reported in Europe and in most soybean regions of the United States and Canada. During the early 1950s it was prevalent in the north central United States, causing losses of up to 50%. First observed in the southern United States in 1973, stem canker is now considered to be endemic throughout the area. The causal fungus can be highly virulent in that area, causing 100% losses in susceptible cultivars.

Symptoms

The first symptoms occur during the early reproductive stages, with the appearance of small, reddish brown lesions, usually near a leaf node (Fig. 49). As the season progresses, the lesions expand longitudinally to form dark brown to black, sunken cankers. Leaf symptoms develop at this stage, with interveinal chlorosis and necrosis characteristic of diseases that restrict water conduction in the plant. Occasionally a necrosis of the terminal meristem develops, with characteristic shepherd's-crook curling.

Stem canker induces a lower frequency of multiple infections and less longitudinal development of cankers in northern regions than in southern regions. Both northern and southern stem canker cause plant death as a result of stem girdling, with toxins at least partially responsible for symptom production and plant death. Disease severity varies greatly from year to

Fig. 49. Lesions and cankers associated with nodes of a soybean plant infected by *Diaporthe phaseolorum* var. *caulivora*, the causal fungus of stem canker. (Courtesy P. A. Backman)

year, even within the same location planted with the same susceptible cultivar. Environment during in the early vegetative stages greatly influences disease severity.

Causal Organism

The causal fungus is *Diaporthe phaseolorum* (Cke. & Ell.) Sacc. var. *caulivora* Athow & Caldwell (anamorph rare). Sufficient differences may exist between the fungi implicated in northern and southern stem canker to warrant separation at the level of variety or forma specialis.

Perithecia may form on infected tissue late in the growing season but are more commonly observed on overwintering stems. In northern stem canker, they are black and globose and are formed in caespitose groups of two to 12. In southern stem canker, perithecia are almost always borne singly. The northern strain produces perithecia embedded in cortical tissues. They measure $165–340 \times 282–412\,\mu m$, with a protruding beak, which is highly variable in length and width and can be very long. The eight-spored asci ($30–40 \times 4–7\,\mu m$) are sessile and elongate-clavate, with thin, evanescent walls, which are slightly thickened at the base. Ascospores ($8–12 \times 3–4\,\mu m$) are hyaline, elongate-ellipsoidal, two-celled, slightly constricted at the septum, and biguttulate in each cell.

In cultures grown on potato-dextrose agar, colonies of the northern fungus are white and closely appressed, later becoming flocculent, whereas those of the southern type are tan to buff, with early development of tufted aerial mycelium, which later becomes lanose. Chlamydospores form in southern isolates, giving a tan to buff coloration to the colonies. Stromata in northern isolates are circular (less than 2 mm in diameter) and randomly distributed. Those in southern isolates are irregular in shape (2–10 mm in diameter) and occasionally fuse to form a large stroma. Perithecia readily form; the neck width of those from northern isolates is half that of perithecia from southern isolates. Pycnidia rarely form on potato-dextrose agar but have been reported more frequently on other media. Southern isolates contain only alpha conidia. Both forms of the fungus are homothallic.

Epidemiology

Primary spread to previously unaffected areas comes about through the movement of infested debris, contaminated equipment, and infested seed. Numerous reports indicate that *D. phaseolorum* var. *caulivora* is found on seed. In southern stem canker, the level of infected seed has not been found above 2%; reports of the northern disease indicate much higher levels of seed infection. Although there is good circumstantial evidence that the disease can be spread by infected seed, there is no direct evidence of seed transmission. In infested fields the source of primary inoculum for new plantings is from overwintering debris.

Perithecia can develop anytime after senescence or plant death but require 8 or more days of free moisture and temperatures above $20°C$. In the southern United States these conditions are most often met in the late spring or early summer. Mature perithecia produce viable ascospores for a period of about 3 weeks. Soybeans develop maximal levels of the disease if infected at growth stage V_3. Disease losses are progressively less if cultivars are infected at later growth stages. Only the most susceptible cultivars become diseased when infected in the late vegetative stages.

An unusual aspect of stem canker is that all cultivars are colonized by the pathogen, but only susceptible cultivars allow the disease to develop. Once infection occurs, the stem canker fungus can produce perithecia in all cultivars, but the quantities of spores produced are related to the level of susceptibility of the cultivar.

Pycnidia of *D. phaseolorum* var. *caulivora* have been observed in the field (particularly in the southern United States), but they have not yet been shown to play a role in the epidemiology of stem canker.

The northern and southern fungi have different growth responses to temperature. This may be in part responsible for the regional distribution of these two biotypes.

Control

Northern stem canker was controlled in the early 1950s by removing the highly susceptible cultivars Hawkeye and Blackhawk from production. Typically, the northern disease is now found only in isolated outbreaks that do not cause severe losses except in individual fields. The causal fungus is endemic across the northern states, as evidenced by its frequent isolation from seeds. Southern stem canker has presented a far more difficult problem. In the southern United States at least 90% of the cultivars have some degree of susceptibility (Table 7). On the basis of recommendations developed in the early 1980s, all highly susceptible cultivars were removed from production, but many of the other cultivars retained because of other agronomic characteristics still suffer significant losses. Cultivars resistant to stem canker are almost always susceptible to the soybean cyst nematode.

Further reductions in disease severity are accomplished through rotation.

Fungicidal spray programs coordinated with infection periods have been quite successful in reducing disease severity.

Table 7. Relative Resistance of Selected Soybean Cultivars to Southern Isolates of *Diaporthe phaseolorum* var. *caulivora*, the Causal Organism of Stem Canker[a]

| Resistant | | Moderately Resistant | | Moderately Susceptible | | Susceptible | |
Maturity Group[b]	Cultivar	Maturity Group	Cultivar	Maturity Group	Cultivar	Maturity Group	Cultivar
I	Blackhawk	V	Terra Vig 505	II	Gnome	V	AP 55
	Hardin		Deltapine 105	III	Elf		A 5539
	Hodgson 78		Deltapine 345		Sprite		RA 502
II	Century		Wilstar 550	IV	Pixie	VI	RA 604
	Corsoy 79		Shiloh	V	Bedford		Brysoy 9
	Hawkeye	VI	Centennial		Forrest		Bradley
III	Cumberland		Davis		Essex	VII	Coker 237
	Williams 82		RA 680		A 5474		Bragg
V	Bay		Coker 156	VI	Jeff		McNair 770
VI	Tracy-M	VII	Wright		Lee 74		Wilstar 790
	Shiloh		Ransom		S69-96		RA 701
VII	Braxton		Coker 317		Deltapine 506	VIII	Coker 338
VIII	Dowling		GaSoy 17	VII	Gregg		Hutton
		VIII	Coker 368		Gordon		RA 801
			Coker 488	VIII	Foster	IX	Jupiter R
			Cobb		Kirby		Santa Rosa R

[a] Source: Backman et al, 1985.
[b] Reactions of cultivars in maturity groups I, II, III, IV, and IX are based on greenhouse inoculation tests; all others are based on field tests.

Benzimidazole fungicides are best suited to control *D. phaseolorum* var. *caulivora*.

A model relating losses at harvest to the time of infection, the growth stage at the time of infection, and the susceptibility of cultivars was developed at Auburn University.

Selected References

Athow, K. L., and Caldwell, R. M. 1954. A comparative study of Diaporthe stem canker and pod and stem blight of soybean. Phytopathology 44:319-325.

Backman, P. A., Weaver, D. B., and Morgan-Jones, G. 1985. Soybean stem canker: An emerging disease problem. Plant Dis. 69:641-647.

Keeling, B. L. 1985. Soybean cultivar reactions to soybean stem canker caused by *Diaporthe phaseolorum* var. *caulivora* and pathogenic variation among isolates. Plant Dis. 69:132-133.

Keeling, B. L. 1988. Influence of temperature on growth and pathogenicity of geographic isolates of *Diaporthe phaseolorum* var. *caulivora*. Plant Dis. 72:220-222.

McGee, D. C. 1983. Epidemiology of soybean seed decay by *Phomopsis* and *Diaporthe* spp. Seed Sci. Technol. 11:719-729.

McGee, D. C., and Biddle, J. A. 1987. Seedborne *Diaporthe phaseolorum* var. *caulivora* in Iowa and its relationship to soybean stem canker in the southern United States. Plant Dis. 71:620-622.

Ploetz, R. C., and Shokes, F. M. 1985. Soybean stem canker incited by ascospores and conidia of the fungus causing the disease in the southeastern United States. Plant Dis. 69:990-992.

Smith, E. F., and Backman, P. A. 1988. Soybean stem canker: An overview. Pages 47-55 in: Soybean Diseases of the North Central Region. T. D. Wyllie and D. H. Scott, eds. American Phytopathological Society, St. Paul, MN.

(Prepared by P. A. Backman, D. C. McGee,
and G. Morgan-Jones)

Pythium Rot

Several species of *Pythium*—*P. aphanidermatum* (Edson) Fitzp., *P. debaryanum* Hesse, *P. irregulare* Buis., *P. myriotylum* Drechs., and *P. ultimum* Trow—have been reported to cause seed decay, pre- and postemergence damping-off (Fig. 50), and root rot of soybeans. The disease complex is found wherever soybeans are grown. All five species are cosmopolitan in soils and attack a wide range of crop plants.

Pythium rot can occur in soybean plants at any time from seed germination to mid-season but is primarily a seedling disease. In general, only scattered individual plants or small groups of plants are killed, and although stands may be reduced, the disease usually does not cause economic losses.

Symptoms

Seedlings infected with *P. debaryanum* or *P. ultimum* may develop different symptoms. Wet rot symptoms develop in seedlings infected with *P. ultimum*, whereas baldhead symptoms (retarded development of the growing point) are prominent in seedlings infected with *P. debaryanum*. In cold, wet soils, both species can cause seed rot and preemergence damping-off.

Seedlings infected with *P. ultimum* may fail to emerge. If they do emerge, the hypocotyl is often attacked at the crook region (the upper hypocotyl) or at the growing root tip. Seedlings attacked at the crook region generally die; those attacked at the root tip initiate secondary roots and generally survive.

Freshly invaded stem tissues are translucent; older lesions become brown, and a watery soft rot is produced. Later, if dry weather sets in, the plants appear dry and shredded (Fig. 50). Cortical tissues in the larger roots may disintegrate and slough off, exposing the central woody cylinder. Smaller roots may decay or break away when plants are pulled from the soil.

Seedlings infected with *P. debaryanum* develop small, black, dry, sunken lesions on the cotyledons, especially on the upper surface. These lesions may coalesce. Water-soaking does not occur. The apical meristem may be severely stunted. The hypocotyl may be two to three times larger in diameter than normal and may remain free of lesions or any discoloration at or near the soil line. The swelling of the hypocotyl may be confused with herbicide injury (see Herbicide Damage). Two axillary buds may develop at the cotyledonal node. Stems from these buds grow slowly, with shorter internodes and much smaller trifoliolate leaves than normal.

Little information is available regarding the symptoms produced by the other *Pythium* spp. In warm, moist soils, *P. aphanidermatum* causes a root rot closely resembling that caused by other *Pythium* spp.

Infection by *Pythium* spp. is often followed by infection by other microorganisms, which can mask typical symptoms.

Causal Organisms

All five *Pythium* spp. that attack soybeans are easy to isolate from infected tissues and grow readily on various sugar-rich media. *Pythium* spp. produce asexual structures (sporangia) on coenocytic hyphae. Sporangia of different species differ in size and shape. They germinate either by germ tube or by producing zoospores. The biciliate zoospore, after encystment, forms a germ tube. Oogonia are spherical and smooth-walled. Antheridia differ from species to species in size, shape, origin, and number per oogonium.

P. aphanidermatum. Oogonia are smooth, spherical, and usually terminal. Antheridia (one or two per oogonium) are monoclinous or diclinous, typically intercalary, and dome-shaped. Oospores are smooth, aplerotic, single, and moderately thick-walled. Sporangia are inflated or filamentous and are branched or unbranched; they are produced freely. Zoospores are formed. Hyphae are 2–8 μm thick.

P. debaryanum. Oogonia are smooth, terminal or intercalary, and usually spherical. Antheridia (one to six per oogonium) are monoclinous or diclinous. Oospores are smooth and aplerotic; they germinate directly by a germ tube that, at low temperatures, gives rise to a vesicle in which zoospores are formed. The few sporangia are spherical to oval and terminal or intercalary; they germinate by germ tube or zoospores. Hyphae are usually 5 μm in diameter.

P. ultimum. Oogonia are smooth, spherical, and mostly terminal. Antheridia (usually one per oogonium) are monoclinous and arise first below the oogonium. Oospores are aplerotic, single, spherical, smooth, and thick-walled; they germinate by a germ tube that, at low temperatures, emits a vesicle in which zoospores are formed. Sporangia are mostly

Fig. 50. Soybean seedlings wilting and dying in the field from Pythium root rot, caused by *Pythium* spp. (Courtesy H. J. Walters)

terminal and spherical and germinate by one or more germ tubes, thus functioning as conidia. Intercalary, barrel-shaped sporangia are formed. They germinate directly by germ tube; zoospores are rarely formed. Hyphae are 1.7–6.5 μm thick.

Disease Cycle and Epidemiology

P. aphanidermatum, P. debaryanum, and *P. ultimum* are soil inhabitants and subsist as saprophytes, colonizing crop residues. They overseason as oospores.

Low soil temperatures (10–15°C) are most favorable for damping-off caused by *P. debaryanum* and *P. ultimum.* Damping-off decreases at soil temperatures above 15°C, and the number of damped-off seedlings declines sharply at temperatures above 22°C. Emergence of seedlings is greatly reduced when soybean seeds are planted in soils infested with *P. debaryanum* and *P. ultimum* and exposed to cold (4°C); the longer the exposure continues, the fewer seedlings emerge. Seedlings up to 10 days old are more susceptible to damping-off than older plants.

Infection by *P. aphanidermatum* occurs between 25 and 36°C. Hyphae enter the host directly through the epidermis and spread intercellularly within the cortex and endodermis, ultimately invading the stele. In seeds, hyphae penetrate the seed coat and spread extensively in the cotyledons.

Pythium infection commonly occurs at high soil moistures. Under these conditions, not only is the oxygen content of the soil low, but the water hinders uptake of oxygen by soybean seeds. Oxygen deficiency is most harmful during the initial stage of seed germination. Even a short period of oxygen deficiency changes the kinds and amounts of exudates from germinating soybean seeds so that they are richer in sugars and amino acids than in a normal soil environment. These exudates stimulate growth of *P. ultimum* and probably other *Pythium* spp.

Control

1. Use a resistant cultivar if available and if the disease persists year after year. Several cultivars resistant to Pythium root rot have been reported.

2. Plant good-quality seeds free of cracks and capable of at least 85% germination in warm (above 18°C), well-drained, fertile soil that is well prepared. Where possible, plow under weeds or cover crops several weeks before planting.

3. Avoid excessive irrigation the first 10 days after planting.

4. Use a seed protectant fungicide.

5. Metalaxyl-containing fungicides may be applied in the seed furrow.

Selected References

Brown, G. E., and Kennedy, B. W. 1966. Effect of oxygen concentration on Pythium seed rot of soybean. Phytopathology 56:407-411.

Hildebrand, A. A., and Koch, L. W. 1952. Observations on a root and stem rot of soybeans new to Ontario, caused by *Pythium ultimum* Trow. Sci. Agric. 32:574-580.

Lehman, S. G., and Wolf, F. A. 1926. Pythium root rot of soybeans. J. Agric. Res. 33:375-380.

Rayside, P. A., Mitchell, D. J., and Gallaher, R. N. 1982. Populations of *Pythium* spp. and associated microorganisms in a minimum tillage–multicropping system. (Abstr.) Phytopathology 72:996.

Schlub, R. L., and Lockwood, J. L. 1981. Etiology and epidemiology of seedling rot of soybean by *Pythium ultimum*. Phytopathology 71:134-138.

Simcox, K. D., and Paxton, J. D. 1984. Induced pathogenicity in glyphosate-treated soybeans. (Abstr.) Phytopathology 74:1271.

Thomson, T. B., Athow, K. L., and Laviolette, F. A. 1971. The effect of temperature on the pathogenicity of *Pythium aphanidermatum, P. debaryanum,* and *P. ultimum* on soybean. Phytopathology 61:933-935.

Red Crown Rot

Red crown rot, also known as Cylindrocladium root rot, black root rot, and Calonectria root rot, has been reported in Brazil, Cameroon, eastern Europe, Japan, Korea, and areas of the southeastern United States. Losses range from negligible to more than 50% of plants damaged or killed. The causal fungus can cause serious diseases of peanut (*Arachis hypogaea* L.) and small grains. Isolates from peanut are pathogenic to soybeans; those from soybeans are pathogenic to common bean (*Phaseolus vulgaris* L.) and other legume crops.

Symptoms

Symptoms usually appear in the early pod stage (R_3–R_4), with yellowing of the tops of individual plants. Later, interveinal tissues of many leaves turn light brown. Wilting and defoliation may occur. Infected plants are scattered or confined to restricted areas in affected fields.

The causal fungus invades roots, eventually rotting the entire root system. Tissues within the stalk become grayish brown, 8–10 cm above the soil line. Reddish orange perithecia normally develop on the stem from the soil surface to 8 cm above the surface (Plate 11). Stem tissues are sometimes reddish even in the absence of perithecia.

Causal Organism

Calonectria crotalariae (Loos) Bell & Sobers (anamorph *Cylindrocladium crotalariae* (Loos) Bell & Sobers) is the pathogen. Soybeans also are susceptible to *Cylindrocladium clavatum* C. S. Hodges & May, *Cylindrocladium floridanum* Sobers & Seymour, and *Cylindrocladium scoparium* Morgan.

Superficial, reddish orange perithecia of *Calonectria crotalariae* develop on the lower stem at an advanced stage of root and lower stem decay (Fig. 51A and B and Plate 11). The perithecia contain clavate asci (95–138 × 13–19 μm), bearing curved, hyaline, mostly three-septate ascospores (34–58 × 6.3–7.8 μm) (Fig. 51D). Conidia (58–107 × 4.8–7.1 μm) are cylindrical, straight, hyaline, and three-septate (Fig. 51C). Microsclerotia are highly variable, with dimensions in the range of 33–311 × 22–133 μm.

Disease Cycle and Epidemiology

Microsclerotia are the primary survival and dispersal propagules. The exact roles of ascospores and conidia are not known. On peanut, short-range spread of the fungus occurs mostly by the movement of microsclerotia in host debris. Spread by ascospores has also been reported. Dispersal of microsclerotia by moving water has been suggested.

Control

1. Use resistant cultivars where available.
2. Delay planting to reduce disease severity.

Selected References

Berner, D. K., Berggren, G. T., Pace, M. E., White, E. P., Gershey, J. S., Freedman, J. A., and Snow, J. P. 1986. Red crown rot: Now a major disease of soybeans. La. Agric. 29:4-5, 24.

Black, M. C., and Beute, M. K. 1984. Effects of rotations with susceptible and resistant peanuts, soybeans, and corn on inoculum efficiency of *Cylindrocladium crotalariae* on peanuts. Plant Dis. 68:401-405.

Dianese, J. C., Ribeiro, W. R. C., and Urben, A. F. 1986. Root rot of soybean caused by *Cylindrocladium clavatum* in central Brazil. Plant Dis. 70:977-980.

Fortnum, B. A., and Lewis, S. A. 1983. Effects of growth regulators and nematodes on Cylindrocladium black root rot of soybean. Plant Dis. 67:282-284.

Phipps, P. M., and Beute, M. K. 1979. Population dynamics of *Cylindrocladium crotalariae* microsclerotia in naturally-infested soil. Phytopathology 69:240-243.

Sung, J. M., and Chung, K. W. 1983. Yield loss associated with disease severity of soybean black root rot by *Cylindrocladium crotalariae*. Korean J. Mycol. 11:99-101.

(Prepared by G. T. Berggren, Jr., and J. P. Snow)

Rhizoctonia Diseases

Rhizoctonia diseases, including pre- and postemergence damping-off, root and stem decay, and leaf and bud blight, have been reported in all soybean-growing areas of the world (see Rhizoctonia Aerial Blight). Pre- and postemergence damping-off and root decay can reduce stands by as much as 50%, and yield losses of up to 40% have been recorded in Brazil and the United States (Fig. 52). The causal fungus has a wide host range, which includes most field crops, vegetables, ornamentals, and fruits.

Symptoms

Rhizoctonia preemergence blight occurs immediately after the plumule emerges from the seed. The sprouted seed is killed and decayed by the causal fungus. Damping-off can occur a few days after emergence. Lesions appear at the base of the seedling stem and on roots just below the soil line. These may enlarge into a sunken lesion, which may girdle the stem (Fig. 53).

Young lesions are brown, dark brown, or reddish, depending on the isolate and soil conditions. Destruction of seedling tissues may in part be due to toxins produced by the fungus. Seedlings that survive the initial infection may develop a reddish brown cortical decay above the crown (Fig. 53). The discoloration may extend into the pith of the stem and large roots. Lesions or cankers may so weaken the stem that plants break off in mid-season or die. Decay may continue intermittently throughout the growing season, with continuing death of plants. Infected plants that survive show some yield reduction. Nodulation (see Beneficial Bacteria) may be significantly reduced.

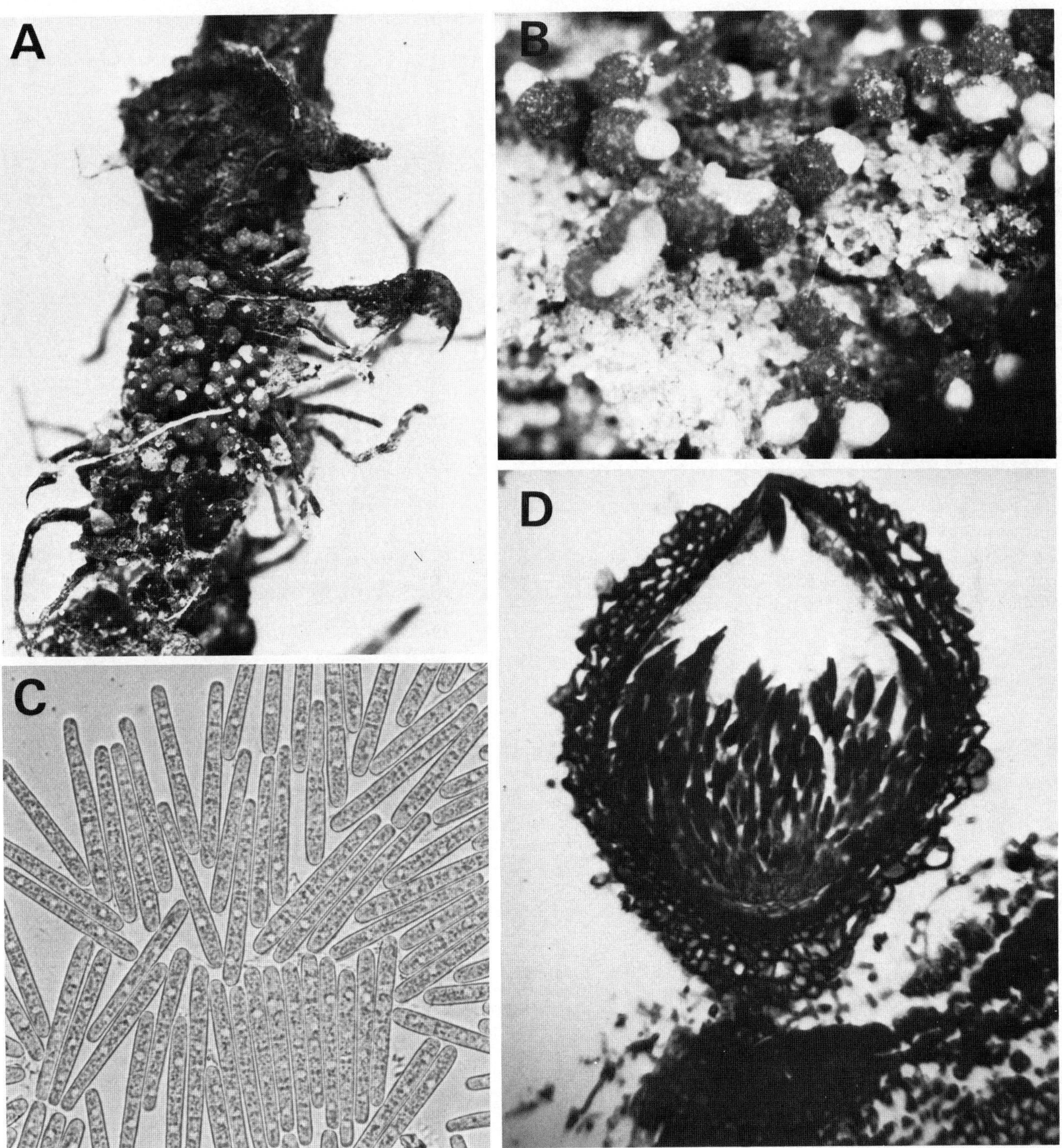

Fig. 51. *Calonectria crotalariae*, the causal fungus of red crown rot. **A,** Perithecia being discharged on the stem of a peanut plant (*Arachis* sp.). **B,** Perithecia. **C,** Conidia. **D,** Cross section of a mature perithecium, with asci and ascospores. (Courtesy R. C. Rowe [A and B] and M. K. Beute [C and D])

Causal Organism

Rhizoctonia solani Kühn is the causal fungus; its teleomorph is *Thanatephorus cucumeris* (Frank) Donk (basionym *Hypochnus cucumeris* Frank; syns. *H. solani* Prill. & Delacr. and *H. filamentosus* Pat.). *Rhizoctonia* spp. are ecologically specific and differ morphologically. *R. solani* is highly variable in cultural characteristics, pathogenicity, and responses to environmental changes. Isolates that cause root and stem decay of soybeans may not cause aerial and foliage blight (see Rhizoctonia Aerial Blight). Isolates of *R. solani* are classified by anastomosis groups (AG), and 13 subgroups belonging to AG-1 through AG-9 have been described. Most soybean pathogens fall into AG-4, but some have been shown experimentally to be in AG-1, AG-2-I, AG-2-II, AG-3, and AG-5. Several ungrouped isolates are pathogenic to soybeans.

Most isolates grow well on potato-dextrose agar at 25–30°C and also grow on a variety of other liquid and solid media. Colony zonation may be characteristic of some isolates. Young mycelium is hyaline; older colonies vary from white to shades of brown and are powdery or hairy. Aerial hyphae, when present, become yellow and then brown with age. Typical hyphae are thick; their size and color vary with age. Cell diameter ranges from 4 to 12 μm (averaging 7.5 μm), and the ratio of length to width of cells generally exceeds 5:1. Hyphal cells are multinucleate (Fig. 54), the number of nuclei varying among AGs and isolates. Young hyphae branch in the direction of growth and are constricted at the point of union with the mother hypha (Fig. 54). Dolipore septa are characteristic of this fungus and are easily seen in the branch after the constriction (Fig. 54). Mature hyphae branch at right angles, but the branch later bends in the direction of growth. Branching at acute angles is common. Hyphae may produce simple or branched chains or moniliform cells, which are hyaline to brown, barrel-shaped, pyriform, irregular, or lobate, and about three times as long as they are wide.

Sclerotia are not produced by all isolates. When present, they consist of compact masses of moniliform cells, are typically superficial on the host, and are chestnut brown to black. Sclerotia have been reported in the pith of infected plants in Florida.

Epidemiology

R. solani is primarily a soil inhabitant and has excellent saprophytic ability. It survives up to 3 months in dry culture and 9 months in liquid culture. The extent of saprophytism varies among isolates. The fungus can overseason in the absence of host tissue, grows well in soil, and colonizes all types of plant debris. Growth in soil depends on nutrient supply; soil moisture, temperature, and pH; and competition from other soil microorganisms. The population of the fungus is distributed mainly in the upper 10 cm of the soil, decreasing with depth to about 50 cm below the surface. When environmental conditions are optimal for the fungus, disease severity is directly related to inoculum potential. An inoculum density of 0.1 μg of mycelia per gram of soil can cause disease in soybeans. The fungus produces detectable pectolytic and proteolytic enzymes.

Control

1. Use fungicide seed protectants. Foliar application of systemic fungicides appears promising.
2. Maintain good soil drainage.
3. Use less susceptible cultivars.

Selected References

Anderson, N. A. 1982. The genetics and pathology of *Rhizoctonia solani*. Annu. Rev. Phytopathol. 20:329-347.

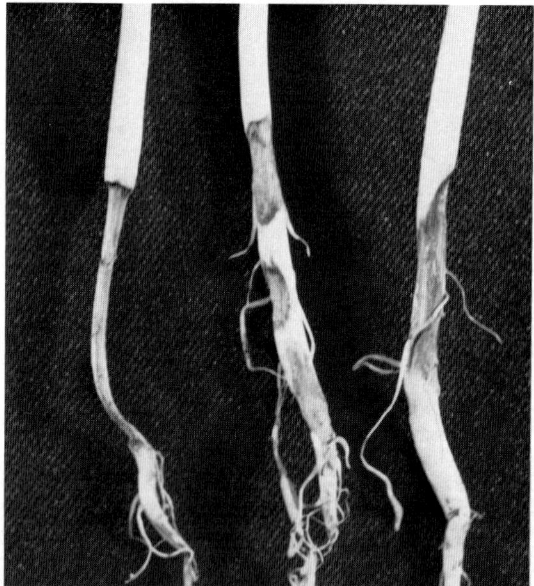

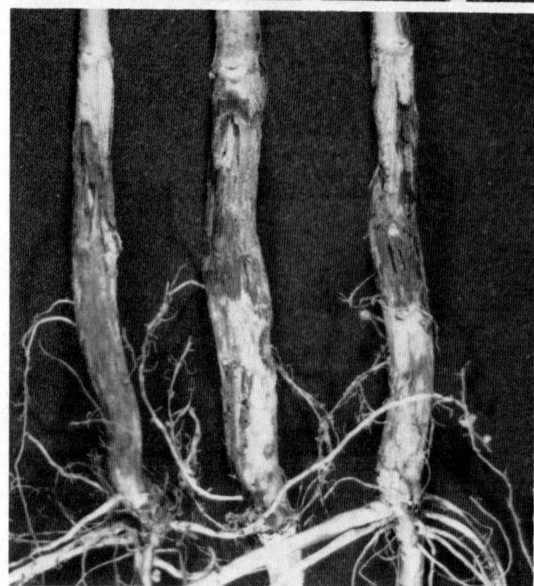

Fig. 53. Soybeans infected with *Rhizoctonia solani*, causing a seedling root decay (above), which may result in postemergence damping-off, and a cortical decay above the crown (below), which weakens and kills maturing plants. (Courtesy Z. Liu)

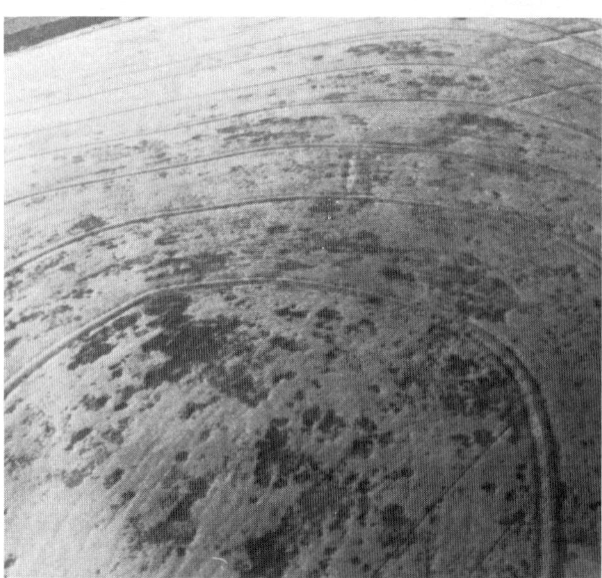

Fig. 52. Aerial view of a soybean field in Brazil with patches of dead plants killed by *Rhizoctonia solani* alone or in combination with other fungi. (Courtesy J. T. Yorinori)

Bolkan, H. A., and Ribeiro, W. R. C. 1985. Anastomosis groups and pathogenicity of *Rhizoctonia solani* isolates from Brazil. Plant Dis. 69:599-601.

Parmeter, J. R., Jr., ed. 1970. *Rhizoctonia solani*: Biology and Pathology. University of California Press, Berkeley. 255 pp.

Tachibana, H., Jowett, D., and Fehr, W. R. 1971. Determination of losses in soybeans caused by *Rhizoctonia solani*. Phytopathology 61:1444-1446.

(Prepared by Z. Liu)

Sclerotinia Stem Rot

Sclerotinia stem rot of soybeans was first reported in Hungary in 1924 and in the United States in 1946. It has since been reported in Argentina, Brazil, Canada, India, Nepal, South Africa, and other soybean-growing areas. The disease is of minor importance in the United States except for local outbreaks, in which plants can be killed before maturity. Yield loss has been estimated at 0.25 Mg/ha for each 10% increment of diseased plants. Each year an estimated 15% loss occurs in central and southern Brazil.

One hazard associated with this disease is the contamination of soybean seeds with sclerotia during harvesting. A small center of infection within a field may thus contaminate a large volume of seeds. At a foreign port of entry, even a few sclerotia in a shipment intended for human consumption are grounds for rejection. Further, sclerotia can germinate if free water is present, resulting in seed decay problems in storage.

The causal fungus has a wide host range, which includes numerous field crops, vegetables, ornamentals, and fruits.

Symptoms

Wilting and eventual death of upper leaves, the first symptoms of Sclerotinia stem rot, generally develop in growth stages R_2 and R_3. Leaves become grayish green as necrosis begins, and they turn brown in time. They remain attached to the stems. Foliar symptoms of Sclerotinia stem rot could be mistaken for late-season Phytophthora root rot, brown stem rot, and stem canker. Sites of infection originate at stem nodes and usually develop 10–50 cm above the soil line. Stem infections are rarely observed at the soil line. Initially, water-soaked lesions develop at nodes, change from tan to white, and spread acropetally and basipetally from nodes. At crop maturity, diseased stem tissues have a shredded appearance if disturbed and are white, with a reddish discoloration frequently interspersed within diseased stem tissues and at the borders of lesions. Multiple stem infections also occur. Lesions eventually girdle the stem, thus inhibiting the movement of water and nutrients to the upper foliage. Side branches and pods may also be infected. Pod development and pod fill above stem lesions are greatly reduced. Cottony mycelial growth on all diseased plant parts is a characteristic sign of the pathogen.

Large, black, round to irregularly shaped sclerotia of varying size form on stems, which are partially covered with dense white mycelium. Sclerotia are often formed in the stem pith and are conspicuous only when the stem is opened (Fig. 55). Occasionally, they are found in pods.

Seeds may become infected within diseased pods. If infected early, seeds are flattened and shriveled and sometimes replaced by black sclerotia. Pods above a stem lesion are reduced in number and size or are absent in many situations.

Causal Organism

The causal fungus, *Sclerotinia sclerotiorum* (Lib.) d By. (syn. *Whetzelinia sclerotiorum* (Lib.) Korf & Dumont), produces hard, black, irregularly shaped sclerotia (2–20 mm in diameter).

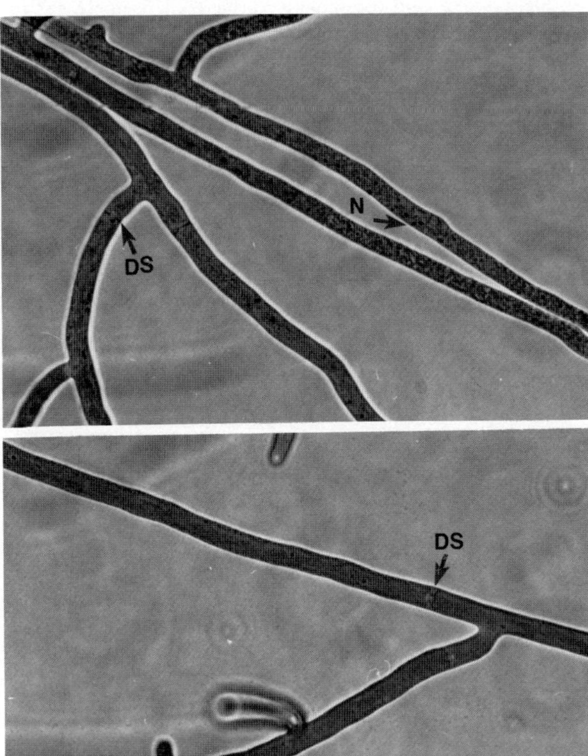

Fig. 54. Stained hyphae of *Rhizoctonia solani*, in a bright-field photomicrograph showing right-angled branching and constriction at the point of union with the mother hypha, dolipore septa (DS), and nucleoli (N) (×280). (Courtesy Z. Liu)

Fig. 55. Sclerotia of *Sclerotinia sclerotiorum* (left); sclerotia developing on the surface (center) and in the pith (right) of soybean stems with Sclerotinia stem decay. (Courtesy M. C. Shurtleff)

They are many times larger than those of *Macrophomina phaseolina* (Tassi) Goid., the charcoal rot fungus, and are easily distinguished from the smaller, round, tan to reddish brown sclerotia of *Sclerotium rolfsii* Sacc.

A sclerotium may produce one to many cup-shaped apothecia, borne on slender stalks. The apothecia are 0.5–2 mm or more in diameter, funnel-shaped to discoid, and light tan to brown. The hymenium is composed of closely packed, eight-spored asci interspersed with slender, simple, hyaline paraphyses. The asci (81–252 × 4–22 μm) are narrow and cylindrical or cylindrical-clavate.

S. sclerotiorum grows well on potato-dextrose agar within a range of 1–32°C (the optimum is 24°C). The fungus is highly variable in culture, producing white to chocolate brown aerial mycelium and black sclerotia, which vary in size and shape (Fig. 55). Spherical microconidia are formed in culture but rarely germinate in water or on nutrient agar. Apothecia and ascospores develop in alternating light and darkness at 12–15°C, provided the sclerotia are incubated in a water-saturated medium.

Disease Cycle and Epidemiology

Sclerotia of *S. sclerotiorum* survive in the soil for long periods. They are highly resistant to many fungicides, to dry heat up to 70°C, and to prolonged freezing and thawing. Those within 5 cm of the soil surface germinate in the field by producing apothecia. Prolonged periods of low soil temperatures (5–15°C) and high soil moisture (−0.25 bar for 10–14 days) are most favorable for apothecial development.

Ascospores are forcibly ejected from asci and are windborne to nearby stems, branches, and pods, where infection occurs if the relative humidity is high. Blossoms are colonized and serve as a nutrient base for infection of stems. Infection of leaves, petioles, and internodes can occur through contact with adjacent diseased plants or through wounds.

Studies to determine the optimal environmental conditions that favor infection of soybeans by *S. sclerotiorum* are lacking. Most observers suggest that stem rot is most likely to develop during extended periods of cool, moist weather. However, the disease has been observed during periods of low rainfall when the crop was planted with row spacings of less than 38 cm. The occurrence of Sclerotinia stem rot is favored by the inclusion of another host crop in rotation. Broadleaf weeds may serve as hosts and thereby as a source of inoculum. Seed contaminated with sclerotia is the most likely means of introducing the pathogen into previously uninfested fields.

Soybean cultivars range from moderately resistant to very susceptible to *S. sclerotiorum*, and differences in cultivar reactions can also be due to plant architecture, maturity, and lodging characteristics. The expression of resistance in soybeans is affected by cultural practices that modify the canopy environment, such as reduced row width, plant population, and irrigation applied at flowering.

Control

1. Avoid planting soybeans directly after common bean (*Phaseolus* spp.), sunflower, and other susceptible crops.

2. Avoid planting soybeans with row spacings of less than 76 cm in fields with a history of Sclerotinia stem rot.

3. Avoid planting cultivars that lodge, and avoid management practices that promote lodging or result in a dense canopy.

4. Avoid excessive irrigation until flowering has ceased.

5. Soybean cultivars differ in susceptibility to Sclerotinia stem rot, and cultivar selection can reduce risks associated with specific management systems.

Selected References

Bedi, K. S. 1961. Factors affecting the viability of sclerotia of *Sclerotinia sclerotiorum* (Lib.) deBary. Indian J. Agric. Sci. 31:236-245.

Boland, G. J., and Hall, R. 1988. Epidemiology of Sclerotinia stem rot of soybean in Ontario. Phytopathology 78:1241-1245.

Chamberlain, D. W. 1951. Sclerotinia stem rot of soybeans. Plant Dis. Rep. 35:490-491.

Chun, D., Kao, L. B., Lockwood, J. L., and Isleib, T. G. 1987. Laboratory and field assessment of resistance in soybean to stem rot caused by *Sclerotinia sclerotiorum*. Plant Dis. 71:811-815.

Grau, C. R. 1988. Sclerotinia stem rot of soybean. Pages 56-66 in: Soybean Diseases of the North Central Region. T. D. Wyllie and D. H. Scott, eds. American Phytopathological Society, St. Paul, MN.

Thompson, A. H., and van der Westhuizer, G. C. A. 1979. *Sclerotinia sclerotiorum* (Lib.) deBary on soybean in South Africa. Phytophylactica 11:145-148.

(Prepared by C. R. Grau)

Sclerotium Blight

Sclerotium blight, also known as southern blight, southern stem rot, and white mold, is a minor disease of soybeans. Significant yield losses can occur in soybean monoculture or short rotations of soybeans with other crops susceptible to the disease. First described in tomato in Florida in 1893 by P. H. Rolfs, Sclerotium blight has subsequently been reported in 500 species in 100 plant families. It is usually found but not restricted to the tropics and warmer regions worldwide.

Symptoms

Soybeans are susceptible to Sclerotium blight anytime from seedling emergence through pod fill. Generally, the disease develops on isolated plants scattered throughout a field. Under favorable conditions, spread to adjacent plants in a row can be rapid. Infection usually occurs at or just below the soil surface. Light brown lesions, which quickly darken, enlarge until the hypocotyl or stem is girdled. A sudden yellowing or wilting of plants is usually the first symptom. Leaves of infected plants turn brown, dry, and often cling to the dead stem. A leaf spot phase is characterized by circular, tan to brown, zonate lesions with dark brown margins. Twin-stem abnormality is associated with seedling infection.

The most characteristic sign of the Sclerotium blight pathogen is the white mat of fungal mycelia forming on stem bases, leaf debris, and the soil surface around infected plants. The mycelial mat may extend several centimeters up the stem above the soil line. Hot, humid weather favors the development of the mycelial mat, which commonly disappears during drought. Numerous tan to brown, spherical sclerotia, about the size of mustard seed, form on infested plant material (Plate 12) and on the soil surface.

Causal Organism

The causal fungus, *Sclerotium rolfsii* Sacc. (teleomorph *Athelia rolfsii* (Curzi) Tu & Kimbr.), produces cordlike, septate, hyaline hyphae, 5–9 μm in diameter, with regular clamp connections. Hyphal branching typically occurs at right angles to the point of origin. Asexual spores are not produced. The sexual stage rarely develops on crop debris or culture media. Clavate basidia, bearing hyaline, pyriform basidiospores (1.0–1.7 × 6–12 μm), are produced on an exposed hymenium. Spherical sclerotia, 1–2 mm in diameter, develop 6 to 12 days after formation of the mycelial mat. The outer covering, or rind, of sclerotia is initially white and later turns dark brown, but the center remains white.

Disease Cycle

Sclerotium blight generally occurs in hot, wet weather. Maximum disease development occurs at temperatures close to the 25–35°C optimum for mycelial growth and sclerotial germination. Damage is favored by high moisture levels both in soil and under a dense crop canopy. Drought conditions often precede severe disease outbreaks. Sclerotium blight is most common and damaging on well-aerated sandy or sandy loam soils. Disease incidence may be reduced by the application of calcium and nitrogen fertilizers.

S. rolfsii survives from one season to the next by producing sclerotia. Factors such as temperature, moisture, proximity of sclerotia to a susceptible host, and their depth in the soil influence their survival rate. Sclerotia on or near the soil surface survive longer than those buried in the soil. They are readily dispersed with soil moved by water, wind, or tillage equipment. Seed transmission has been reported but is much less important than transmission by soilborne inoculum.

Germination of sclerotia in water-saturated soil is stimulated by volatile organic exudates from crop residue. Since sclerotia are deficient in energy, the fungus gains energy by colonizing undecomposed organic matter before attacking a susceptible host. The fungus, by producing enzymes that degrade oxalic acid, pectin, and cellulose, kills host tissue in advance of its invading mycelia.

Control

1. Alternating soybeans or other susceptible crops with nonhost crops such as maize, grain sorghum, or pasture grasses or clean fallowing for two years may prevent a buildup of inoculum to damaging levels. In fields heavily infested with *S. rolfsii*, a three- to four-year rotation with nonhost crops may be needed.

2. Burying crop debris and sclerotia to a depth of 15 to 25 cm with a moldboard plow may reduce inoculum carry-over and delay disease development.

3. Movement of soil during cultivation around soybean stems may increase disease.

4. Planting cultivars tolerant to Sclerotium blight can reduce disease damage. The cultivars FFR 666, Hood, Jackson, Palmetto, and Shelby are tolerant. Other cultivars showing some tolerance are Hill, Scott, Date, Pickett, Semmes, Tracy, Bragg, and Hardee.

Selected References

Aycock, R. 1966. Stem rot and other diseases caused by *Sclerotium rolfsii*. N.C. Agric. Exp. Stn., Tech. Bull. 174. 202 pp.

Beute, M. K., and Rodríguez-Kábana, R. 1979. Effect of volatile compounds from remoistened plant tissues on growth and germination of sclerotia of *Sclerotium rolfsii*. Phytopathology 69:802-805.

Beute, M. K., and Rodríguez-Kábana, R. 1981. Effects of soil moisture, temperature, and field environment on survival of *Sclerotium rolfsii* in Alabama and North Carolina. Phytopathology 71:1293-1296.

Punja, Z. K., Carter, J. D., Campbell, G. M., and Rossell, E. L. 1986. Effects of calcium and nitrogen fertilizers, fungicides, and tillage practices on incidence of *Sclerotium rolfsii* on processing carrots. Plant Dis. 70:819-824.

Rolfs, P. H. 1893. The tomato and some of its diseases. Fla. Agric. Exp. Stn., Bull. 21. 38 pp.

Shew, B. B., and Beute, M. K. 1984. Effects of crop management on the epidemiology of southern stem rot of peanut. Phytopathology 74:530-535.

(Prepared by W. S. Gazaway and A. K. Hagan)

Thielaviopsis Root Rot

Thielaviopsis root rot, also called black root rot, was first reported in Canada in 1925 and was later reported in Germany, the Soviet Union, and the United States. The disease is widespread in Michigan and in southwestern Ontario and may occur in other northern soybean-growing areas.

Symptoms

A dark brown to black necrosis of the cortex of hypocotyls, taproots, and fibrous roots forms below the soil line, sometimes resulting in severe root rot (Fig. 56). The causal fungus frequently severs the taproot 2.5–5 cm below the soil surface. The soybean plant responds by forming shallow adventitious roots, which support the plant when soil moisture is ample, but which cannot sustain it under drought conditions.

The disease is most severe early in the season when soils are cool (16–20°C). Diseased seedlings may be killed. Surviving older plants may wilt under drought stress.

Limited seedborne infection has been reported.

Thielaviopsis root rot often occurs together with Phytophthora root rot.

Causal Organism

Thielaviopsis basicola (Berk. & Br.) Ferr. (syn. *Thielavia basicola* Zopf) is primarily a soil inhabitant. Mycelium is almost hyaline. Conidia are formed within short, hyaline conidiophores swollen at the base and tapering toward the apex. Numerous cylindrical conidia are borne within a conidiophore tip and are pushed out slowly in chains.

The fungus produces characteristic chlamydospores within infected tissues (Fig. 57). They form on short, clavate, lateral hyphae in a series of short, brown, thick-walled cells, which soon separate. Chlamydospores survive for long periods in the soil, whereas conidia are short-lived.

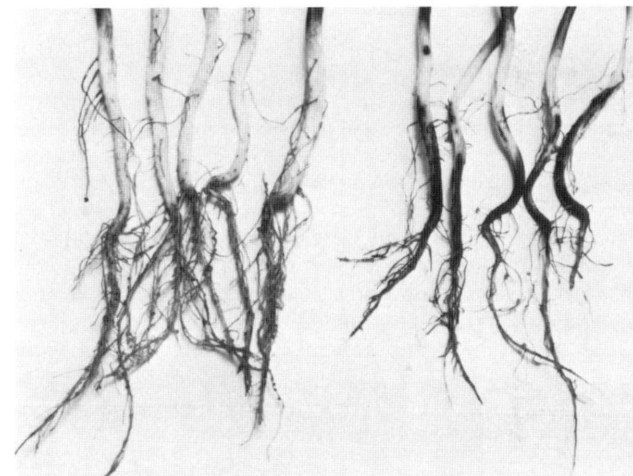

Fig. 56. Soybean plants with symptoms of Thielaviopsis root rot, caused by *Thielaviopsis basicola*. (Courtesy J. L. Lockwood)

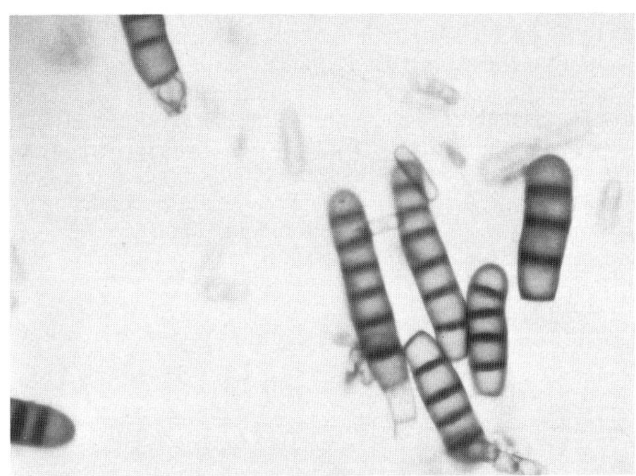

Fig. 57. Chlamydospore chains of *Thielaviopsis basicola*, the causal fungus of Thielaviopsis root rot. In soil, the chlamydospores separate into individual cells. Conidia are visible in the background. (Courtesy J. L. Lockwood)

T. basicola is not readily isolated from soil or diseased plant tissues. It can be selectively isolated on fresh carrot disks or on specific selective agar media.

Control

1. Resistant cultivars in laboratory tests indicate that commercial cultivars vary considerably in susceptibility.

2. Avoid the herbicide chloramben in soils where *T. basicola* is present. Greenhouse tests and limited field tests have shown that chloramben increases the severity of root rot caused by the fungus.

Selected References

Anderson, T. R. 1984. Thielaviopsis root rot in Ontario and susceptibility of commercial cultivars to root inoculation. Can. J. Plant Pathol. 6:71-74.

Lee, M., and Lockwood, J. L. 1977. Enhanced severity of Thielaviopsis basicola root rot induced in soybean by the herbicide chloramben. Phytopathology 67:1360-1367.

Specht, L. P., and Griffin, G. J. 1985. A selective medium for enumerating low populations of *Thielaviopsis basicola* in tobacco field soils. Can. J. Plant Pathol. 7:438-441.

(Prepared by J. L. Lockwood)

Other Fungi
Associated with Root Rots

Gliocladium roseum

Gliocladium roseum Bainier, a common soilborne fungus, is frequently found associated with soybean roots but is not known to be a pathogen of soybeans. *Gliocladium* is identical in growth habit to *Verticillium*. Verticillate conidiophores are present in immature colonies and branch repeatedly as the colony ages to form penicillate structures with phialides all present on the same level. It produces conidia in mucilaginous drops, which may fuse to form larger balls of conidia enveloped in slime. The verticillate conidiophores in immature colonies can lead to the misidentification of *Gliocladium* as *Verticillium*.

Selected References

Mueller, J. D., and Sinclair, J. B. 1986. Occurrence and role of *Gliocladium roseum* in field-grown soybeans in Illinois. Trans. Br. Mycol. Soc. 86:297-301.

Schiller, C. T., and Sinclair, J. B. 1984. Microorganisms associated with soybean vascular exudates and plant parts. Int. J. Plant Pathol. 2:1-4.

Mycoleptodiscus terrestris

Mycoleptodiscus terrestris (Gerd.) Ostazeski causes a crown and root rot of soybeans in central and southern Illinois and other soybean-growing areas of the United States and in India and North Africa. Distinct symptoms are difficult to characterize, because the pathogen is usually associated with other root rot fungi in the field. *M. terrestris* causes a postemergence damping-off of soybean seedlings, characterized by a reddish brown to black cortical decay of crown and root tissues. It also causes a dark decay of the lateral root system and taproot of older plants, similar to that produced by *Rhizoctonia solani* Kühn (see *Rhizoctonia* Diseases).

Selected Reference

Gray, L. E. 1978. Mycoleptodiscus terrestris root rot of soybeans. Plant Dis. Rep. 62:72-73.

Neocosmospora vasinfecta

Neocosmospora vasinfecta E. F. Smith causes a dark discoloration of soybean stems and pith, with occasional chlorosis and defoliation of the lower leaves. The symptoms may be confused with those caused by *Phialophora gregata* (Allington & Chamberlain) W. Gams (see Brown Stem Rot). Bright-field microscopy reveals mycelium in pith and xylem vessels of stem sections of young plants infected with *P. gregata*, whereas plants infected with *N. vasinfecta* have mycelium in pith tissues only.

Selected Reference

Gray, F. A., Rodríguez-Kábana, R., and Adams, J. R. 1980. Neocosmospora stem rot of soybeans in Alabama. Plant Dis. 64:321-322.

Beneficial Fungi

The roots of soybeans can be infected with vesicular-arbuscular endomycorrhizal fungi, which form a symbionic relationship beneficial to the plant. Feeder roots are transformed into morphological structures called mycorrhizae. Endomycorrhizal fungi can be found associated with soybeans wherever the crop is grown. The fungal hyphae invade the cortical tissues of feeder roots and then form haustoria (arbuscules) or vesicles (large, swollen hyphae) within the cortical cells. It is difficult to distinguish a colonized root from one without the fungus. A loose mycelial growth appears on the external surface of invaded roots and may aid in the uptake of nitrogen, phosphorus, and water.

Selected References

Anderson, A. J. 1988. Mycorrhizae—Host specificity and recognition. Phytopathology 78:375-378.

Graham, J. H. 1988. Interactions of mycorrhizal fungi with soilborne plant pathogens and other organisms: An introduction. Phytopathology 78:365-366.

Linderman, R. G. 1988. Mycorrhizal interactions with the rhizosphere microflora: The mycorrhizosphere effect. Phytopathology 78:366-371.

Smith, G. S. 1988. The role of phosphorus nutrition in interactions of vesicular-arbuscular mycorrhizal fungi with soilborne nematodes and fungi. Phytopathology 78:371-374.

Virus Diseases

More than 600 viruses—macromolecules composed of either ribonucleic acid (RNA) or deoxyribonucleic acid (DNA) surrounded by a protective protein or lipoprotein coat—are known to infect plants. Virus particles, or virions, can multiply only within living cells and thus are obligate parasites. Virions may be filamentous rods (long and either rigid or flexuous), isometric particles (roughly spherical), or bacilliform particles (elliptical). They range in diameter from 10 to 70 nm, and rod lengths can exceed 2 μm. Observation of individual virus particles requires a high-resolution electron microscope.

Color Plates

1. Bacterial blight (right) and bacterial pustule (left). (Courtesy CNPSoja)

2. Bacterial blight. (Courtesy B. W. Kennedy)

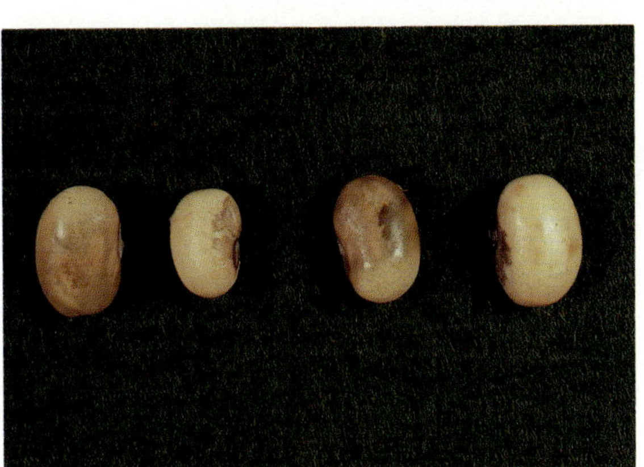

3. Bacterial pustule. (Courtesy B. W. Kennedy)

4. Bacterial wildfire. (Courtesy CNPSoja)

5. Anthracnose. (Courtesy P. R. Hepperly)

6. Rust on the lower surface of a leaflet. (Courtesy C. C. Yeh)

7. Rust lesions: type RB (below) and type TAN (above). (Courtesy C. C. Yeh)

8. Phytophthora stem decay. (Courtesy M. Ferguson)

9. Soybean pods with pycnidia (center) and stromata (right) of *Diaporthe* (*Phomopsis*) and a healthy pod (left). (Courtesy P. R. Hepperly)

10. *Phomopsis*-infected seeds (left) and healthy seeds (right). (Courtesy P. R. Hepperly)

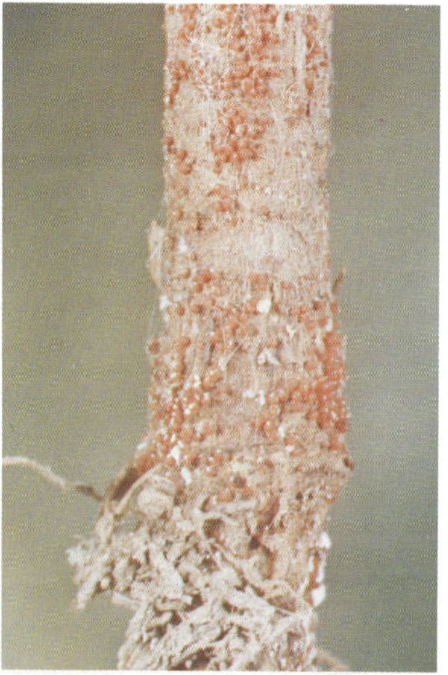

11. Red crown rot. (Courtesy F. H. Smith)

12. Sclerotium blight. (Courtesy Clemson University)

13. Bean pod mottle. (Courtesy J. P. Fulton)

14. Cowpea chlorotic mottle. (Courtesy C. W. Kuhn)

15. Peanut mottle. (Courtesy C. W. Kuhn)

16. Leaflets with common mosaic (center) and yellow mosaic (right) and a healthy leaflet (left). (Courtesy Illinois Agricultural Experiment Station)

17. Common mosaic. (Courtesy M. C. Shurtleff)

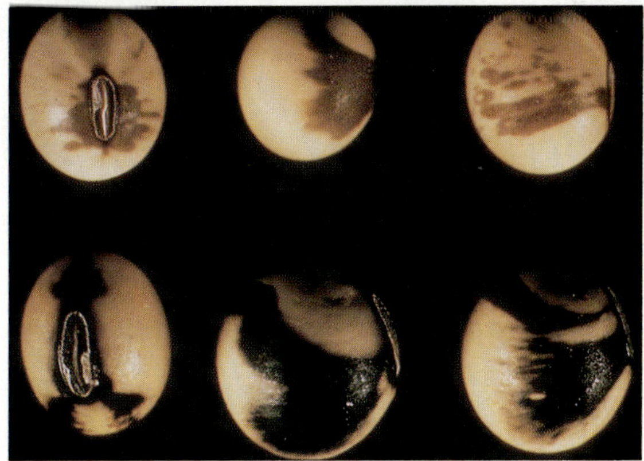

18. Common mosaic seed discoloration. (Courtesy J. H. Johnson)

19. Seed with triple infection by *Cercospora sojina* (left), soybean mosaic virus (center), and *C. kikuchii* (top). (Courtesy J.T.Yorinori)

20. Bud blight, caused by tobacco ringspot virus, with shepherd's-crook and dark pith symptoms. (Courtesy Illinois Agricultural Experiment Station)

21. Bud blight, caused by tobacco ringspot virus. (Courtesy B. J. Jacobsen)

22. Bud blight with pod lesions, caused by tobacco ringspot virus. (Courtesy M. C. Shurtleff)

23. Brazilian bud blight, caused by tobacco streak virus. (Courtesy A. S. Costa)

24. Yellow mosaic. (Courtesy M. C. Shurtleff)

25. Yellow mosaic in India. (Courtesy J. B. Sinclair)

26. Abutilon mosaic of soybeans. (Courtesy A. S. Costa)

27. Soybean cyst nematode injury. (Courtesy M. C. Shurtleff)

28. Soil peds containing cysts of the soybean cyst nematode. (Courtesy G. R. Noel)

29. Lesion nematode injury to roots. (Courtesy N. Acosta V.)

30. Sporulation of *Cercospora sojina* (left) and *C. kikuchii* (right) on germinating seeds. (Courtesy J. T. Yorinori)

31. Seeds with purple seed stain (left) and healthy seeds (right). (Courtesy J. T. Yorinori)

32. Sudden death syndrome. (Courtesy J. C. Rupe)

33. Sudden death syndrome, with chlorotic areas coalescing in interveinal streaks. (Courtesy J. C. Rupe)

34. Soil crusting injury. (Courtesy W. O. Scott)

35. Frost injury. (Courtesy W. O. Scott)

36. Frost injury to soybean leaflets. (Courtesy H. G. Johnson)

37. Hail injury. (Courtesy W. O. Scott)

38. Heat canker. (Courtesy L. E. Sweets)

39. Lightning injury. (Courtesy A. S. Costa)

40. Sunburn injury. (Courtesy W. O. Scott)

41. Boron toxicity. (Courtesy L. F. Welch)

42. Calcium deficiency. (Courtesy M. C. Shurtleff)

43. Iron deficiency. (Courtesy N. R. Usherwood)

44. Excess magnesium (left) and magnesium deficiency (right). (Courtesy M. C. Shurtleff)

45. Manganese deficiency. (Courtesy M. C. Shurtleff)

46. Manganese toxicity. (Courtesy N. R. Usherwood)

47. Nitrogen deficiency. (Courtesy L. F. Welch)

48. Potassium deficiency. (Courtesy L. F. Welch)

49. Potassium deficiency, with necrosis of chlorotic areas of leaflets. (Courtesy H. G. Johnson)

50. Zinc deficiency. (Courtesy K. Onki)

51. Chloramben injury. (Courtesy E. L. Knake)

52. Dicamba injury. (Courtesy E. L. Knake)

53. Dicamba injury to soybean leaflets. (Courtesy E. L. Knake)

54. Trifluralin injury. (Courtesy E. L. Knake)

55. Linuron injury. (Courtesy E. L. Knake)

56. Metribuzin injury resulting from use of the herbicide with an organophosphate insecticide. (Courtesy D. E. Kuhlman)

57. Atrazine injury. (Courtesy M. C. Shurtleff)

58. 2,4-D drift injury and 2,4-DB direct injury. (Courtesy E. L. Knake)

59. 2,4-DB injury. (Courtesy E. L. Knake)

60. Vernolate injury. (Courtesy E. L. Knake)

61. Carbaryl injury. (Courtesy R. K. Howell)

62. Dicofol injury. (Courtesy R. K. Howell)

63. Chlorine injury. (Courtesy J. B. Sinclair)

64. Ammonia injury. (Courtesy R. K. Howell)

65. Ozone injury. (Courtesy R. K. Howell)

66. Sulfur dioxide injury. (Courtesy R. K. Howell)

Viruses are transmitted to plants through wounds created by animal vectors (mainly arthropods and nematodes) or fungal vectors; by parasitic plants; by mechanical inoculation; by pollination; and by deliberate or accidental human activities, such as planting infected seed and propagating plants from infected cuttings or by grafting or budding infected plant materials.

Viral symptoms range from latent infections to plant death. Viruses may induce stunting, rugosity, mosaic patterns, yellowing or reddening of foliage, and necrosis. The action of two or more viruses in a single plant (multiple infection) can be additive, synergistic, or cross-protective (one strain preventing infection by another strain of the same virus).

Viruses are identified by particle morphology; biochemical, physical, and serological properties; mode of transmission; and host specificity. The types of symptoms of indicator plant hosts also aid in identification. Most plant viruses are named for the original or major host attacked and the symptoms produced. Plant viruses are grouped taxonomically by particle morphology, type and size of nucleic acid, genome structure, mode of transmission, and serological properties.

At present, identifying virus and viruslike diseases in the field is more an art or a game of chance than a science. Indistinct symptomology, the occurrence of several viruses in the same area, the existence of virus strains, the presence of singly and doubly infected plants, and inadequate symptom characterizations confound disease diagnosis in the field. Moreover, symptoms of virus infection may be confused with those caused by mycoplasmas or mycoplasmalike organisms (see Mycoplasmalike Diseases). Satisfactory clinical techniques, however, are available to identify many viruses. Genetic and nutritional abnormalities may be ruled out if a causal agent is transmitted from diseased to healthy plants; electron microscopy of diseased tissue reveals virus particles that are absent in healthy plants; serological tests detect relationships with known plant viruses; and physiochemical tests such as electrophoresis of viral double-stranded RNAs and complementary DNA probes to detect viral nucleic acids are also being used successfully.

Soybeans are known to be susceptible to at least 111 viruses or virus strains (Table 8). Several virus diseases of soybeans are not known to occur in nature but have been induced under experimental conditions. Viruses that infect soybeans may contain single-stranded RNA, single-stranded DNA, or double-stranded DNA. At this time, 33 viruses known to occur naturally in soybeans are believed to have some economic significance; 18 of them of particular importance belong to 10 of the 27 plant virus groups.

Selected References

Gibbs, A., and Harrison, B. 1976. Plant Virology: The Principles. John Wiley & Sons, New York. 292 pp.

Harrison, B. D., and Murant, A. F., eds. 1977. Descriptions of Plant Viruses, with supplements. Commonwealth Mycological Institute and Association of Applied Biologists, Kew, Surrey, England.

Kado, C. I., and Agarwal, H. O. 1972. Principles and Techniques in Plant Virology. Van Nostrand-Reinhold, London. 688 pp.

Matthews, R. E. F. 1981. Plant Virology, 2nd ed. Academic Press, New York. 897 pp.

Matthews, R. E. F. 1982. Classification and nomenclature of viruses. Third report of the International Committee on Taxonomy of Viruses. Intervirology 12:131-296.

Smith, K. M. 1972. A Textbook of Plant Virus Diseases, 3rd ed. Academic Press, New York. 684 pp.

Tisselli, O., Sinclair, J. B., and Hymowitz, T. 1980. Sources of Resistance to Selected Fungal, Bacterial, Viral and Nematode Diseases of Soybeans. INTSOY Ser. 18. College of Agriculture, University of Illinois at Urbana-Champaign. 134 pp.

Walkey, D. G. A. 1985. Applied Plant Virology. John Wiley & Sons, New York. 329 pp.

(Prepared by J. W. Demski, C. W. Kuhn, and J. B. Sinclair)

Bean Pod Mottle Virus

Bean pod mottle virus (BPMV) causes a mottling disease of soybeans, which was first noted in Arkansas, North Carolina, and Virginia in 1958 and has since been detected in Nebraska.

Table 8. Plant Viruses Reported to Occur in Soybeans Naturally or by Inoculation[a]

Rod-shaped or filamentous particles	Isometric particles (continued)
Carlaviruses	Luteoviruses
Cowpea mild mottle	African soybean dwarf
Pea streak	Bean leaf roll
Red clover vein mosaic	Beet western yellows
Potexviruses	Chick-pea stunt
Clover yellow mosaic	Indonesian soybean dwarf
Foxtail mosaic	Legume yellows
Pea wilt	Milk vetch dwarf
White clover mosaic	Pea leaf roll
Potyviruses	Peanut rosette (assistor)
Adzuki bean mosaic	Soybean dwarf
Bean common mosaic	Subterranean clover red leaf
Bean yellow mosaic	Nepoviruses
Blackeye cowpea mosaic	Arabis mosaic
Clover yellow vein	Cacao necrosis
Cowpea aphid-borne mosaic	Cherry leaf roll
Passion fruit woodiness	Raspberry ringspot
Peanut chlorotic ring mottle	Tobacco ringspot
Peanut green mosaic	Tomato black ring
Peanut mottle	Tomato ringspot
Peanut stripe	Pea enation mosaic virus
Soybean crinkle	Pea enation mosaic
Soybean mosaic	Sobemoviruses
Soybean yellow bud	Rice yellow mosaic
Sugarcane mosaic	Southern bean mosaic
Watermelon mosaic 2	Subterranean clover mottle
Yellow mosaic	Tobacco necrosis virus
Tobamoviruses	Bean stipple streak
Peanut clump	Tobacco necrosis
Soybean yellow vein	Tomato spotted wilt virus
Sunn hemp mosaic	Tomato spotted wilt
Tobacco mosaic	Tombusviruses
Tobraviruses	Glycine mottle
Pea early-browning	Tephrosia symptomless
Soybean fleck	Tymoviruses
Tobacco rattle	Clitoria yellow vein
Isometric particles	Kennedya yellow mosaic
Bromoviruses	Ononis yellow mosaic
Bean yellow stipple	Plantago mottle
Broad bean mottle	Geminiviruses
Cowpea chlorotic mottle	Bean golden mosaic
Caulimoviruses	Horse gram yellow mosaic
Peanut chlorotic streak	Mung bean yellow mosaic
Soybean chlorotic mottle	Soybean crinkle leaf
Comoviruses	Tobacco leaf curl
Bean pod mottle	Bacilliform particles
Broad bean stain	Alfalfa mosaic
Broad bean wilt	Ungrouped viruses
Chick-pea mosaic	Abutilon mosaic
Cowpea mosaic	Bean chlorosis
Cowpea mottle	Bean chlorotic ringspot
Cowpea severe mosaic	Bean mild mosaic
Glycine mosaic	Bean severe mosaic
Quail pea mosaic	Bean yellow dot
Red clover mottle	Bean yellow stripe
Cucumoviruses	Black gram mottle
Cucumber mosaic	Catjang mosaic
Peanut stunt	Groundnut rosette
Soybean stunt	Guar top necrosis
Tomato aspermy	Lupine viruses A, B, C
Dianthoviruses	Michigan alfalfa
Carnation ringspot	Pea stunt
Red clover necrotic mosaic	Pepper mosaic
Ilarviruses	Soybean mild mosaic
Tobacco streak	Subterranean clover mosaic
	Sunflower ringspot mosaic
	Wisconsin pea streak

[a] Groupings based on Harrison and Murant, 1977.

The disease may be found in doubly infected plants in association with soybean mosaic virus (SMV). The combination causes severe losses: BPMV alone may reduce yields by 10–17%, but in association with SMV, losses may exceed 60%. Maximum losses occur when plants are infected at the seedling stage. The disease decreases pod formation and reduces seed size, weight, and number.

Symptoms

Symptoms are most obvious during periods of rapid plant growth and cool conditions. Young leaves in the upper canopy exhibit green to yellow mottling (Fig. 58 and Plate 13). Under cool greenhouse conditions, mechanically inoculated leaves develop chlorotic mottling. Symptoms are masked during periods of high temperatures and are not observed on plants after pod set.

Leaves of infected plants under water stress are less turgid than leaves of healthy plants. Water stress of infected plants is more detrimental in the reproductive stages than in the vegetative stage. Pod formation may be reduced by as much as 40% if infected plants are under water stress in the reproductive stages.

Plants infected with BPMV may have green stems after the pods have matured. Such plants may retain petioles after leaf blades have abscised.

BPMV alone can cause necrosis of new terminal leaf growth. However, it does not cause the stem tip to curve or become extremely brittle, as it does in combination with SMV and as tobacco ringspot virus, tobacco streak virus, and SMV can do singly. BPMV and SMV interact synergistically in mixed infections in soybeans to cause more severe symptoms and greater yield losses than single infections (see Soybean Mosaic Virus).

Symptom severity is related to the infectivity of viral strains. As the period of infection lengthens, the specific infectivity of the virus in leaves decreases.

Seeds from plants inoculated with BPMV have a higher incidence of infection by seedborne fungi.

Causal Agent

BPMV is a member of the comoviruses, which have a bipartite genome. The virions are small isometric particles, about 30 nm in diameter. Purified virus preparations contain three centrifugal components: top RNA-free protein shells (54 S) and two kinds of nucleoprotein particles, middle (91 S) and bottom (112 S). The single-stranded RNAs of the middle and bottom components have molecular weights of 1.5×10^6 and 2.5×10^6, respectively.

Fig. 58. Symptoms caused by bean pod mottle virus on soybean leaves. (Courtesy J. P. Fulton)

BPMV has a thermal inactivation point between 70 and 75°C, a dilution end point of 10^{-4} to 10^{-5}, and longevity in vitro of 62–93 days at 18°C.

The virus is strongly immunogenic, and a variety of serological procedures have been used successfully. BPMV can be detected in soybean seeds and in the bean leaf beetle, *Cerotoma trifurcata* (Forster).

Host Range

The host range of BPMV is limited to the legumes. Besides soybeans and *Glycine soja* Sieb. & Zucc., the virus infects *Desmodium paniculatum* (L.) DC., *Lespedeza cuneata* (Dumont) G. Don, *L. striata* (Thunb.) Hook. & Arn., *L. stipulacea* Maxim., *Mucuna deeringianum* (Bort) Merr., *Phaseolus acutifolius* A. Gray, *P. lunatus* L., *P. vulgaris* L., and *Trifolium incarnatum* L.

Transmission

BPMV is sap-transmissible and has been reported to be seed-transmitted at a frequency of 0.1% in one location. In some fields, the incidence of infected plants can be correlated with the population of the beetle vector *C. trifurcata*. Other beetle vectors include *Colaspis brunnea* (Fabricius), *Colaspis lata* Schaeffer, *Diabrotica balteata* LeConte, *D. undecimpunctata howardi* Barber, and *Epicauta vittata* (Fabricius).

Control

The following cultural practices have been recommended for control of BPMV:

1. Grow at least four rows of maize, forage sorghum, or other tall-growing crops between soybeans and other leguminous crops, such as clovers or alfalfa, and between soybeans and uncultivated areas, such as roadsides, ditch banks, and pastures.

2. Before soybean seedlings emerge, apply a herbicide to kill broadleaf weeds in noncrop areas next to soybean fields.

Selected References

Ghabrial, S. A., and Schultz, F. J. 1983. Serological detection of bean pod mottle virus in bean leaf beetles. Phytopathology 73:480-483.

Lin, M. T., and Hill, J. H. 1983. Bean pod mottle virus: Occurrence in Nebraska and seed transmission in soybeans. Plant Dis. 67:230-233.

Myhre, G. L., Pitre, H. N., Haridasan, M., and Hesketh, L. D. 1973. Effects of bean pod mottle virus on yield components and morphology of soybeans in relation to soil water regimes: A preliminary study. Plant Dis. Rep. 57:1050-1054.

Patel, V. C., and Pitre, H. N. 1971. Transmission of bean pod mottle virus soybean by the striped blister beetle, *Epicauta vittata*. Plant Dis. Rep. 55:628-629.

Ross, J. P. 1969. Effect of time and sequence of inoculation of soybeans with soybean mosaic and bean pod mottle viruses on yields and seed characters. Phytopathology 59:1404-1408.

Schwenk, F. W., and Nickell, C. D. 1980. Soybean green stem caused by bean pod mottle virus. Plant Dis. 64:863-865.

Scott, H. A., van Scyoc, J. V., and van Scyoc, C. E. 1974. Reactions of *Glycine* spp. to bean pod mottle virus. Plant Dis. Rep. 58:191-192.

Skotland, C. B. 1958. Bean pod mottle virus of soybean. Plant Dis. Rep. 42:1155-1156.

Stuckey, R. E., Ghabrial, S. A., and Reicosky, D. A. 1982. Increased incidence of *Phomopsis* sp. in seeds from soybeans infected with bean pod mottle virus. Plant Dis. 66:826-829.

Walters, H. J., and Lee, F. N. 1969. Transmission of bean pod mottle virus from *Desmodium paniculatum* to soybean by the bean leaf beetle. Plant Dis. Rep. 53:411.

(Prepared by J. W. Demski and C. W. Kuhn)

Cowpea Chlorotic Mottle Virus

Cowpea chlorotic mottle virus (CCMV) occurs in cowpea and causes yellow stipple in common bean and mosaic in soybeans. It was first identified in soybeans in 1967. All strains of CCMV infect soybeans, but most research on this virus in soybeans has been conducted with the soybean strain (CCMV-S). CCMV in U.S. soybeans is limited to Georgia, and the bean stipple strain occurs in Cuba. Yield losses caused by CCMV-S can be as high as 37% when plants are inoculated in the seedling stage.

Symptoms

In susceptible soybean cultivars, the predominant symptoms caused by CCMV-S are mosaic (Plate 14) and stunting, which can reduce plant height by up to 42%. Other leaf symptoms include distortion, chlorotic patches, and rugosity. Diseased plants produce fewer seeds, which are of reduced quality. The chemical composition of such seeds is altered.

Causal Agent

CCMV, a member of the bromovirus group, has a tripartite genome. The virions are isometric particles. A single type of polypeptide (180 protein subunits per particle) encapsidates four species of single-stranded, positive-sense RNA, which have molecular weights ranging from about 0.3×10^6 to 1.1×10^6. The genome consists of the three largest RNA molecules.

CCMV has a thermal inactivation point of 65–70°C. Infectivity in vitro is retained for 24 hr at 20°C and for 48 hr at 5°C. Infectivity is retained in dried leaves stored for 3 weeks at 5°C. The dilution end point is between 10^{-4} and 10^{-5}.

Host Range

CCMV causes systemic infection and symptoms in *Vigna cylindrica* (L.) Skeels, *V. unguiculata* (L.) Walpers, and *V. unguiculata* subsp. *unguiculata* cultigroup *sesquipedalis* E. Westphal.

Local lesions are produced on resistant soybean cultivars and *Cassia tora* L., *Chenopodium album* L., *C. hybridum* L., *Cucurbita pepo* L. 'Caserta,' and *Phaseolus acutifolius* A. Gray var. *latifolius* Freeman.

Veinal necrosis occurs in soybean cultivars of intermediate resistance and *Cassia occidentalis* L., *Lablab purpureus* (L.) Sweet, and *Phaseolus vulgaris* L.

Arachis hypogaea L., *Lupinus albus* L., some cultivars of *Nicotiana tabacum* L., *Petunia ×hybrida* Hort. Vilm., *Pisum sativum* L., and *Vicia faba* L. are latent hosts.

Both susceptible soybean and cowpea cultivars are good propagation hosts for CCMV. Assay hosts (developing local lesions) are specific soybean cultivars and *Chenopodium* spp.

Transmission

CCMV-S is sap-transmissible. There is no evidence of seed transmission. Two insects, the bean leaf beetle (*Cerotoma trifurcata* (Forster)) and the spotted cucumber beetle (*Diabrotica undecimpunctata howardi* Barber), are known to transmit CCMV. The bean yellow stipple strain can be transmitted by *D. balteata* LeConte and *C. ruficornis* (Oliver).

Control

Cultivars and plant introductions with the local necrotic lesion type of resistance are common; for example, 347 of 533 genotypes in one test were of this type. Furthermore, other types of resistance (reduced virus concentration and nonsystemic movement) are available and effective under field conditions.

Selected References

Bijaisoradat, M., and Kuhn, C. W. 1985. Nature of resistance in soybean to cowpea chlorotic mottle virus. Phytopathology 75:351-355.

Boerma, H. R., Kuhn, C. W., and Harris, H. B. 1975. Inheritance of resistance to cowpea chlorotic mottle virus (soybean strain) in soybeans. Crop Sci. 15:849-850.

Fernandez, S. R., and Lastres, G. N. 1984. Characteristics and symptoms of bean yellow stripe virus (BYSV) on soybean (*Glycine max*). Aencios Agric. 21:7-11.

Harris, H. B., and Kuhn, C. W. 1971. Influence of cowpea chlorotic mottle virus (soybean strain) on agronomic performance of soybeans. Crop Sci. 11:71-73.

Kuhn, C. W. 1968. Identification and specific infectivity of a soybean strain of cowpea chlorotic mottle virus. Phytopathology 58:1441-1442.

Paguio, O. R., Boerma, H. R., and Kuhn, C. W. 1987. Disease resistance, virus concentration, and agronomic performance of soybean infected with cowpea chlorotic mottle virus. Phytopathology 77:703-707.

Walters, H. J., and Dodd, N. L. 1969. Identification and beetle transmission of an isolate of cowpea chlorotic mottle virus from *Desmodium*. (Abstr.) Phytopathology 59:1055.

(Prepared by J. W. Demski and C. W. Kuhn)

Peanut Mottle Virus

Peanut mottle virus (PMV) causes mottle in soybeans. It was first found occurring naturally in 1971. The disease is common in the southeastern United States and can cause yield reductions of up to 60%. PMV has also been found in soybeans in Australia and Thailand.

Symptoms

The first symptoms of peanut mottle are small, enlarging, chlorotic areas. Dark green islands are produced on young leaves (Plate 15). Later, yellow patches, line patterns, or ring patterns may occur on the third and fourth leaves formed after infection. A mosaic pattern is produced on older leaves. Leaves of infected plants may pucker and curl down at the edges.

Causal Agent

PMV is a typical member of the potyvirus group. Particles are flexuous rods, about 750×12 nm. They contain one molecule of positive-sense, single-stranded RNA that has an approximate molecular weight of 3.0×10^6. The RNA is encapsidated with a single type of capsid protein, each subunit of which has an approximate molecular weight of 35,000.

In sap from infected *Pisum sativum* L., PMV has a thermal inactivation point between 55 and 64°C, the dilution end point is between 10^{-3} and 10^{-4}, and longevity in vitro is 1 to 2 days at room temperature. Significant quantities of highly purified PMV (50–150 µg per gram of infected tissue) can be obtained from diseased *P. sativum*.

The virus is moderately immunogenic and can be evaluated with standard serological tests. Immunosorbent electron microscopy and enzyme-linked immunosorbent assays have demonstrated a serological relationship between PMV and a few other potyviruses. It shares some nucleotide sequence homology with peanut mild virus, peanut stripe virus, and tobacco etch virus.

Variants of PMV have been found in peanut, soybeans, and beans. The variants are distinguished primarily on the basis of host range and disease reactions. Neither serology nor nucleic acid hybridization studies have found definitive differences between the variants.

Host Range

PMV occurs in nature in several important legume crops in addition to soybeans: *Arachis hypogaea* L., *Lupinus albus* L., *L. angustifolius* L., *Phaseolus lunatus* L., *P. vulgaris* L., *Pisum sativum* L., *Trifolium subterraneum* L., *T. vesiculosum* Savi, and *Vigna unguiculata* (L.) Walpers.

Furthermore, the virus has been isolated from a few weed hosts in nature: *Cassia obtusifolia* L., *C. leptocarpa* Benth., *C. occidentalis* L., and *Desmodium canum* (Gmel.) Schinz & Thell.

Several other plant species, mostly legumes, can be infected with PMV: *Calopogonium mucunoides* Desv., *Canavalia ensiformis* (L.) DC., *Cassia bicapsularis* L., *Chenopodium amaranticolor* Coste & Reyn., *Cicer arietinum* L., *Cyamopsis tetragonoloba* (L.) Taub., *Lathyrus odoratus* L., *Macroptilium atropurpureum* (DC.) Urban, *M. lathyroides* (L.) Urban, *Nicotiana benthamiana* Domin, *N. clevelandii* A. Gray, *Phaseolus coccineus* L., *Sesamum indicum* L., *Trifolium hybridum* L., *Trigonella foenum-graecum* L., *Vigna cylindrica* (L.) Skeels, *V. oblongifolia* A. Rich., and *V. unguiculata* (L.) Walpers subsp. *unguiculata* cultigroup *sesquipedalis* E. Westphal.

Transmission

PMV is easily transmitted mechanically to and from soybean plants. In transmission from some infected species, particularly peanut, it is desirable to add antioxidants such as sodium bisulfite and sodium diethyldithiocarbamate to the buffer.

Apparently no seed transmission occurs in soybeans or naturally infected weed hosts. However, seed transmission of PMV has been reported to occur in low frequencies in bean, cowpea, lupine, and peanut.

Several aphid species can readily transmit PMV in a nonpersistent manner: *Aphis craccivora* Koch, *A. gossypii* Glover, *Hyperomyzus lactucae* (L.), *Myzus persicae* (Sulzer), *Rhopalosiphum padi* L., and *R. maidis* (Fitch).

Epidemiology

Peanut seeds infected with PMV appear to provide the source of primary inoculum for infection of susceptible crops. This source has been established for cowpea, peanut, and soybeans. In the United States, approximately 200,000 peanut seeds are planted per hectare. Therefore, a seed transmission frequency as low as 0.1% would provide about two infected seedlings per 100 m^2 in a field. With the presence of aphids, PMV can be spread within the field, and movement to nearby susceptible crops has been demonstrated.

Control

Two control methods should be successful in controlling the soybean disease caused by PMV: isolation from peanut and the use of resistant cultivars. Studies in the United States have shown that very little or no infection develops in susceptible plants that are grown 100 m or more away from infected peanut plants.

Resistant cultivars are available. For example, resistance in Dorman and CNS is conditioned by a single dominant gene, and a single recessive gene controls resistance in Peking.

Selected References

Bays, D. C., Tolin, S. A., and Roane, C. W. 1986. Interactions of peanut mottle virus strains and soybean germ plasm. Phytopathology 76:764-768.

Boerma, H. R., and Kuhn, C. W. 1976. Inheritance of resistance to peanut mottle virus in soybeans. Crop Sci. 16:533-534.

Demski, J. W. 1975. Source and spread of peanut mottle virus in soybean and peanut. Phytopathology 65:917-920.

Kuhn, C. W., Demski, J. W., and Harris, H. B. 1972. Peanut mottle virus in soybeans. Plant Dis. Rep. 56:146-147.

Shipe, E. R., Buss, G. R., and Tolin, S. A. 1979. A second gene for resistance to peanut mottle virus in soybean. Crop Sci. 19:656-658.

Sukorndhaman, M. 1987. Nucleic acid hybridization, serology and host reactions to study classification and detection of peanut mottle virus. Ph.D. diss., University of Georgia, Athens. 92 pp.

(Prepared by J. W. Demski and C. W. Kuhn)

Soybean Dwarf Virus

Soybean dwarf was first described in Japan in 1952. Although disease severity and incidence vary from year to year, the disease is economically important in northern Japan. A six-year survey indicated yield losses depended on cultivar, virus strain, vector density, plant age at the time of infection, and environmental conditions. In years of high incidence, losses of 40% were observed when 50% of the plants were infected. Subterranean clover red leaf virus, now considered a strain of soybean dwarf virus (SDV), causes serious losses in Australia and New Zealand. SDV has not been identified in soybeans in North America, but a virus serologically similar to subterranean clover red leaf virus has been described in California.

Symptoms

Young soybean plants inoculated 1-3 weeks after emergence are most severely affected. Symptom expression ranges from indiscernible to severe, depending on virus strain and cultivar. SDV isolates have been grouped into dwarfing and yellowing types on the basis of host range and symptoms. The strain types can be distinguished in soybean cultivars such as Foster, Hill, Santa Rosa, U7V-1, Asgrow 3127, Bicentennial, Corsoy 79, Cutler 71, Dawson, Douglas, Evans, Hobbit, Hodgson 78, Maple Amber, Pioneer, Sloan, Swift, Wayne 82, Zane, BSR 101, and many of the Tsurunoko types. Plants infected with the dwarfing strain are severely stunted, because of shortened internodes and petioles. As leaflets mature, they curl downward, turn deep green, become thick and brittle, and are smaller than normal (Fig. 59). Plants infected with the yellowing strain are stunted, and a typical interveinal chlorosis develops, increasing in severity as the leaves mature; older leaves become thickened and brittle. Greenhouse-grown plants infected with both strains have rugose leaves and a mixture of symptoms characteristic of the two strains. In the field, the symptoms are usually a mixture of those produced by the dwarfing and yellowing types. Translocation of photosynthates from leaflets to pods is interrupted in infected plants, with the result that necrosis occurs in the vascular bundles. Pod growth and seed number and weight decrease with increased severity of infection. Diseased plants mature late, remaining green until harvest. Mottled seeds have not been observed on infected plants.

Fig. 59. Symptoms caused by soybean dwarf virus: plant and leaf dwarfing (above) and leaf symptoms (below). (Courtesy A. D. Hewings)

Causal Agent

SDV is a member of the luteoviruses. The physiochemical properties of the virus vary slightly from one strain to another. The virions are 26-nm icosahedral particles, which migrate as a single centrifugal component with a buoyant density of about 1.4 in cesium chloride. The major nucleoprotein has a relative molecular mass (M_r) of about 2.5×10^4, and the single-stranded RNA has an M_r of about 1.9×10^6.

SDV is difficult to purify, because the concentration of the virus in the plant is low and extraction from phloem is difficult, usually requiring freezing infected tissue with liquid nitrogen, pulverizing it to a powder, and incubating it with enzymes. The virus can withstand exposure to pectinolytic and cellulolytic enzymes, Triton X-100, and organic solvents such as chloroform, amyl alcohol, and 1-butanol. 1-Butanol appears to reduce the infectivity of purified virions, however. SDV is less stable in cesium salts than in sucrose. Once purified, it remains stable and infectious for months if frozen at -20 or $-80°$C in 0.01 M sodium phosphate buffer at pH 6.0.

SDV is an excellent antigen, and a high-quality polyclonal antiserum can be made in rabbit with as little as 200 μg of the purified virus. Strains of SDV appear to be serologically indistinguishable in double-antibody sandwich enzyme-linked immunosorbent assay using polyclonal antisera. One or more serological techniques have been used to demonstrate that SDV is related to barley yellow dwarf, beet western yellows, and bean leaf roll viruses but not to potato leaf roll virus.

Host Range

In addition to soybeans, SDV infects many genera and species within the Fabaceae. Depending on the virus strain and environmental conditions, symptoms may be very obvious or latent. Legume hosts include *Astragalus sinicus* L. (milk vetch), *Medicago littoralis* Rhode, *M. truncatula* Gaertn., *Phaseolus vulgaris* L. (French bean), *Pisum sativum* L. (pea), *Trifolium alexandrinum* L., *T. dubium* Sibth. (suckling clover), *T. fragiferum* L. (strawberry clover), *T. hybridum* L. (alsike clover), *T. incarnatum* L. (crimson clover), *T. pratense* L. (red clover), *T. repens* L. (white and ladino clovers), *T. subterraneum* L. (subterranean clover), *Vicia faba* L. (broad bean), and *V. sativa* L. (common vetch).

A few nonleguminous hosts—*Beta vulgaris* L. (sugar beet), *Phlox drummondii* Hook., *Spinacia oleracea* L. (spinach), and *Tetragonia tetragonioides* (Pallas) Kuntze—have been identified by enzyme-linked immunosorbent assay, and the identification has been confirmed by back assay with the aphid *Aulacorthum solani* (Kaltenback) to Wayne or Swift soybean indicator hosts.

Transmission

Like all members of the luteovirus group, SDV is persistently transmitted by aphids and causes infections localized in the phloem. Vectors retain the virus through molts and require acquisition and inoculation access periods as long as 48 hr for efficient transmission. SDV has been transmitted by grafting, but no mechanical transmission has been demonstrated with infected plant extracts, and no seed transmission has been demonstrated.

In Japan, six species of aphids were tested as vectors of SDV: *Acyrthosiphon kondoi* Shinji & Kondo, *A. pisum* (Harris), *Aphis craccivora* Koch, *A. glycines* Matsumura, *Aulacorthum solani*, and *Myzus persicae* (Sulzer). Only *A. solani*, the foxglove aphid, was found to transmit the disease. SDV-like isolates in Australia and New Zealand were also transmitted by *A. solani*. In 1980, this species was a major component of the aphid fauna. However, *A. pisum*, the pea aphid, was introduced to Tasmania in 1979, and this change resulted in a shift in isolate type. Most SDV-like isolates recovered from the field in Tasmania are transmitted specifically by *A. pisum*. The virus isolated from legumes in California is transmitted only by this aphid.

Another virus, Indonesian soybean dwarf virus, is vectored in a persistent manner by *A. glycines* and is serologically unrelated to SDV.

Control

Organophosphate insecticides have been used in Japan to control aphids infesting soybean crops. Application into the soil at planting time was found to be the most effective means of aphid control, although topdressing was nearly equal to incorporation. The use of insecticides does not prevent infection by viruliferous winged aphids that migrate into the fields, because they can infect the plants before they die. Insecticides provide satisfactory control of virus spread within a field.

Certain cultivars, such as Adams and Yuuzuru and their F_3 and F_4 progeny, have shown considerable tolerance. A highly resistant cultivar, Tsurukogane, was released in Hokkaido in 1984.

Selected References

Banba, H., Tanimura, Y., and Matsukawa, I. 1986. Breeding for resistance to soybean dwarf virus in soybeans. Trop. Agric. Res. Ser. 19:236-246.

Damsteegt, V. D., and Hewings, A. D. 1986. Comparative transmission of soybean dwarf virus by three geographically diverse populations of *Aulacorthum* (= *Acyrthosiphon*) *solani*. Ann. Appl Biol. 109:453-463.

D'Arcy, C. J., and Hewings, A. D. 1986. Enzyme-linked immunosorbent assays for study of serological relationship and detection of three luteoviruses. Plant Pathol. 35:288-293.

Hewings, A. D., Damsteegt, V. D., and Tolin, S. A. 1986. Purification and some properties of two strains of soybean dwarf virus. Phytopathology 76:759-763.

Matthews, R. E. F. 1981. Plant Virology, 2nd ed. Academic Press, New York. 897 pp.

Rochow, W. F., and Duffus, J. E. 1981. Luteoviruses and yellows diseases. Pages 147-170 in: Handbook of Plant Virus Infections: Comparative Diagnosis. E. Kurstack, ed. Elsevier/North-Holland Biomedical Press, Amsterdam.

Tamada, T. 1975. Studies on the soybean dwarf disease. (In Japanese) Rep. Hokkaido Prefect. Agric. Exp. Stn. 25:1-44.

Tamada, T., and Kojima, M. 1977. Soybean dwarf virus. Descriptions of Plant Viruses, No. 179. Commonwealth Mycological Institute and Association of Applied Biologists, Kew, Surrey, England.

(Prepared by A. D. Hewings)

Soybean Mosaic Virus

Soybean mosaic (also known as soybean crinkle), caused by soybean mosaic virus (SMV), is worldwide in distribution. Other potyviruses closely related to SMV may also be involved. The disease was first observed in the United States in the early 1900s and is believed to have been introduced with the first soybeans brought from Asia. Soybean mosaic is regarded as an important disease in some areas of the world. Yields may be reduced by 50% in any one field. Yield reductions as high as 93% have been recorded in experimentally inoculated plants. Infected plants produce fewer, smaller, lighter-weight seeds, which may have mottled seed coats. Infection by SMV can predispose some cultivars to infection by *Phomopsis longicolla* Hobbs (see Pod and Stem Blight and Phomopsis Seed Decay; Stem Canker; Phomopsis Seed Decay, under Seed Pathology), resulting in a loss of seed germinability.

Symptoms

SMV can be seedborne; infected seeds may fail to germinate or may produce diseased seedlings. Infected seedlings are spindly, with rugose or crinkled unifoliolate leaves, which may be shaped normally but be mottled or may curl longitudinally downward (Fig. 60). Subsequent trifoliolate leaves become prematurely chlorotic and are more severely stunted, mottled, and rugose than the unifoliolate leaves (Fig. 60 and Plates 16 and 17).

A plant's reaction to SMV infection depends on the genotype of the host, the strain of the virus, plant age at the time of infection, and environmental conditions. Plants infected early in the season are stunted, with shortened petioles and internodes. Leaves are reduced in size and generally misshapen and puckered, occasionally with dark green enations along the veins (Plate 17). Affected leaflets are generally asymmetric and curl down at the margins. The youngest and most rapidly growing leaves show the most severe symptoms. Typically, SMV-infected plants mature later than uninfected plants, but infected plants from certain soybean lines mature earlier.

Susceptible cultivars show transitory symptoms 6–14 days after mechanical inoculation. A yellowish veinclearing appears along the small, branching veins of young leaves. Typical rugosity generally appears on the third trifoliolate leaf formed after inoculation. Increasingly more severe symptoms develop on subsequent leaves, which eventually show dark green enations along the main veins (Plate 17). Leaf margins frequently curve down at the sides. Less severe reactions may be limited to mild mosaic symptoms. In some cultivars, plants may develop necrosis. The symptoms associated with necrosis include a brown discoloration of leaf veins; yellowing of leaves; defined, systemic, necrotic lesions on leaves; stunting of the plant; browning of petioles and stems; bud blight (which can be confused with that caused by tobacco ringspot virus); and defoliation, usually leading to plant death.

The severity of symptoms at any given temperature varies from cultivar to cultivar and with the viral strain. Infected plants of different cultivars at 20°C may vary from normal in height to severely stunted, and their leaves may vary from normal in appearance to rugose (Plate 17).

Rugosity is most severe if infected plants are grown at or near 18°C. Symptoms are less severe at 24–25°C and are largely masked above 30°C. Temperature influences the length of time between inoculation and symptom appearance, which ranges from 4 days at 29.5°C to 14 days at 18.5°C.

Expression of pod symptoms depends on cultivar susceptibility, time of infection, viral strain, and temperature. Diseased pods are often stunted and flattened, with less pubescence; they are curved more acutely than pods on healthy plants.

Viable seeds produced in diseased pods may be mottled brown or black (Plates 18 and 19). They are usually smaller than seeds from healthy plants, and germination may be inhibited. No consistent relationship has been found between virus-infected seeds and the location of seeds within a pod or on the plant. (See Seedborne Viruses, under Seed Pathology.)

Bradyrhizobium sp. forms fewer, smaller, and lighter-weight nodules on roots of SMV-infected plants than on healthy plants; the leghemoglobin content is also lower, with the result that nitrogen fixation is reduced. The leghemoglobin content is minimized when plants are inoculated 2 weeks after planting and grown at 21–26°C. Reduction in nodulation is directly related to cultivar susceptibility and day length.

SMV-infected plants in the field may also become infected with bean pod mottle virus (BPMV) (see Bean Pod Mottle Virus). The effect of double infection is synergistic; yield is reduced by as much as 66–80% in doubly infected plants of susceptible cultivars, compared with 8–25% in SMV-infected and 10% in BPMV-infected plants. Severe dwarfing, foliage discoloration, and necrosis occur. Doubly infected foliage appears bronzed in the field. Enations form on the leaves of some doubly infected cultivars. The enations originate from chlorotic areas at the end of the midrib on the upper surface. They are filiform to hirsute, 5–25 mm long, and 1 mm wide at the base, tapering to the tip. They are leaflike, with an upper and a lower epidermis, and sometimes have palisade and spongy parenchyma cell layers. Doubly infected plants have fewer root nodules than singly infected plants. The synergistic effects of double infection on symptoms, yield reduction, seed mottling, and seed transmission of SMV depend to some extent on the strains of the two viruses and on the timing of inoculation.

SMV-infected plant cells, particularly in vascular areas, contain intracellular inclusion bodies that look like pinwheels in cross section and bundles in longitudinal section. They vary in size, generally being about 12–41 μm long. Their number increases between 2 and 3 weeks after inoculation. More inclusions occur in plants infected with a severe strain than a mild one. Both SMV inclusion bodies and BPMV particles may occupy the same cell in doubly infected plants.

Causal Agent

SMV, a member of the potyvirus group, is a flexuous rod averaging about 750 × 15–18 nm (Fig. 61). Virus particles

Fig. 60. Symptoms caused by soybean mosaic virus on cotyledonary leaves of a soybean seedling from an infected seed (above) and a trifoliolate leaf of an inoculated plant (below). (Courtesy G. R. Bowers)

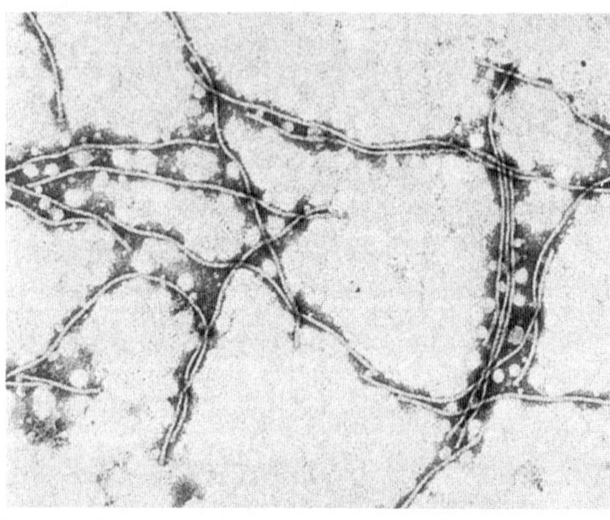

Fig. 61. Soybean mosaic virus particles, in an electron photomicrograph (×14,800). (Courtesy M. Soong and G. M. Milbrath)

ranging from 300 to 900 nm long have been reported; particle size is highly correlated with infectivity, the most infectious particles being over 656 nm long. Several strains of SMV have been recognized, on the basis of the reactions of a differential set of soybean cultivars.

SMV virions have helical symmetry with a pitch of 34°. The nucleic acid in the particles is single-stranded RNA, constituting 5.3% of the particle mass and having a molecular weight of 3.25×10^6.

Thermal inactivation points range from 55 to 70° C. Data on longevity in vitro vary, depending on how the tests were done, from 2 to 5 days. The virus remains infectious in desiccated soybean leaves for 7 days at 25–33° C; it loses infectivity in sap placed 61 cm from an ultraviolet light source for 2 hr. SMV is most stable at pH 6 (near the pH of soybean sap) and loses infectivity at levels below pH 4 and above pH 9.

SMV is difficult to purify, because its particles tend to aggregate. The use of 0.5 M sodium citrate and mercaptoethanol in the early steps of purification minimizes aggregation. Weak borate buffer facilitates redispersal of the pelleted virus after ultracentrifugation. A procedure involving chloroform-butanol clarification, polyethylene glycol precipitation, ultracentrifugation through 30% sucrose, and cesium chloride equilibrium centrifugation gives high yields of the virus while avoiding excessive particle aggregation and host protein contamination.

Host Range

SMV can infect several host species, mostly in the Fabaceae.

It causes local lesions on *Chenopodium album* L., *C. quinoa* Willd., *Cyamopsis tetragonoloba* (L.) Taub., *Dolichos biflorus* L., *Indigofera hirsuta* L., *Lablab purpureus* (L.) Sweet, *Lourea vespertilionis* Desv., *Macroptilium lathyroides* (L.) Urban, *Phaseolus lunatus* L. (developing chlorotic local lesions), some cultivars of *P. vulgaris* L., and *Vigna unguiculata* (L.) Walpers (developing leaf chlorosis).

The virus induces systemic symptoms in soybeans and *Glycine soja* Sieb. & Zucc. and in *Canavalia ensiformis* (L.) DC., *Cassia occidentalis* L., *Crotalaria spectabilis* Roth, *Cyamopsis tetragonoloba* (L.) Taub., *Dolichos falcatus* Klein ex Willd., *Lespedeza stipulacea* Maxim., *L. striata* (Thunb.) Hook. & Arn., *Lupinus albus* L., *L. luteus* L., *Macroptilium lathyroides* (L.) Urban, *Mucuna deeringianum* (Bort) Merr., *Phaseolus lunatus* L. (developing latent to mild infection), *P. nigricans* Haberle ex Martens, some cultivars of *P. vulgaris* L., *Sesbania exaltata* (Raf.) Cory, *Trigonella caerulea* Ser. in DC., and *T. foenum-graecum* L.

Latent infections occur in *Hippocrepis multisiliquosa* L., *Lotus tetragonolobus* L., *Lupinus angustifolius* L., *Phaseolus speciosus* Kunth in H.B.K., some cultivars of *P. vulgaris* L., and *Scorpiurus sulcata* L.

SMV also infects *Amaranthus* sp., *Physalis longifolia* Nutt., *P. virginiana* Mill., *Setaria* sp., and *Solanum carolinense* L.

Transmission

SMV is readily sap-transmitted with or without the use of abrasives. The virus is graft-transmissible but is not transmitted by dodder (*Cuscuta* spp.).

A percentage of seeds from infected plants carry the virus, which may remain viable in seeds for at least 2 years. The extent of seed transmission depends on the virus strain and host genotype. In certain areas that lack vectors that can retain SMV for long periods and hosts in which the virus can overwinter, seed transmission plays an important role in the epidemiology of soybean mosaic.

At least 31 aphid species transmit the virus efficiently in a nonpersistent manner. Aphids become viruliferous by feeding on infected stems, stem tips, and old or young leaves.

Multiplication

Viruses must multiply in infected cells for translocation and systemic infection to take place. SMV moves both up and down in plants and can be detected in all parts of systemically infected plants. Multiplication and movement occur most rapidly at 26° C; no movement is detectable at 10° C. SMV moves slowly from infected cells; 3 days at 26° C and 4 days at 21° C are required for detection of the virus away from the point of inoculation. In systemically infected plants, virus content is correlated with the severity of symptoms.

Control

Resistance to SMV is conditioned by a pair of alleles. Allele *Rsv* (from PI 96983) controls a high level of resistance. Allele *rsv*[+] (from Tokyo via Ogden) controls a lower level of resistance; it is dominant to the susceptible allele *rsv* but recessive to *Rsv*.

Other control measures include the use of SMV-free seeds and roguing of infected plants in fields used to produce planting seeds.

Selected References

Bos, L. 1972. Soybean mosaic virus. Descriptions of Plant Viruses, No. 93. Commonwealth Mycological Institute, Kew, Surrey, England. 4 pp.

Bowers, G. R., Jr., and Goodman, R. M. 1982. New sources of resistance to seed transmission of soybean mosaic virus in soybeans. Crop Sci. 22:155-156.

Cho, E.-K., and Goodman, R. M. 1979. Strains of soybean mosaic virus: Classification based on virulence in resistant soybean cultivars. Phytopathology 69:467-470.

Galvez, G. E. 1963. Host-range, purification, and electron microscopy of soybean mosaic virus. Phytopathology 53:388-393.

Gunasinghe, U. B., Irwin, M. E., and Bernard, R. L. 1986. Effect of a soybean genotype resistant to soybean mosaic virus on transmission-related behavior of aphid vectors. Plant Dis. 70:872-874.

Hill, J. H., Bailey, T. B., Benner, H. I., Tachibana, H., and Durand, D. P. 1987. Soybean mosaic virus: Effects of primary disease incidence on yield and seed quality. Plant Dis. 71:237-239.

Hill, J. H., and Benner, H. I. 1980. Properties of soybean mosaic virus ribonucleic acid. Phytopathology 70:236-239.

Irwin, M. E., and Goodman, R. M. 1981. Ecology and control of soybean mosaic virus. Pages 181-220 in: Plant Diseases and Their Vectors: Ecology and Epidemiology. K. Maramorosch and K. F. Harris, eds. Academic Press, New York.

Kuhl, R. A. S., and Hartwig, E. E. 1979. Inheritance of reaction to soybean mosaic virus in soybean. Crop Sci. 19:372-375.

Lim, S. M. 1985. Resistance to soybean mosaic virus in soybeans. Phytopathology 75:199-201.

Roane, C. W., Tolin, S. A., and Buss, G. R. 1983. Inheritance of reaction to two viruses in the soybean (*Glycine max*) cultivar York crossed with cultivar Lee 68. J. Hered. 74:289-291.

Ross, J. P. 1967. Purification of soybean mosaic virus for antiserum production. Phytopathology 57:465-467.

Ross, J. P. 1969. Effect of time and sequence of inoculation of soybeans with soybean mosaic and bean pod mottle viruses on yields and seed characters. Phytopathology 59:1404-1408.

Soong, M. M., and Milbrath, G. M. 1980. Purification, partial characterization, and serological comparison of soybean mosaic virus and its coat protein. Phytopathology 70:388-391.

(Prepared by J. W. Demski and C. W. Kuhn)

Tobacco Ringspot Virus

Tobacco ringspot virus (TRSV) causes bud blight of soybeans. The disease was first described in the United States in 1941 and has since been reported in Australia, Canada, the People's Republic of China, and the Soviet Union. Of the many diseases caused by TRSV, bud blight of soybeans is the most severe and causes the greatest losses. Yields may be reduced by 25–100%. In general, losses are greatest when young plants are infected or when seeds with a high percentage of TRSV are sown. Yields are also lowered through reduced pod set and seed formation on infected plants. Total protein is increased, and the oil content of seeds from infected plants is decreased.

There is evidence that soybean mosaic virus may be involved in a complex with TRSV.

Symptoms

Plants infected while less than 5 weeks old are stunted. The stunting is not evident when they are grown at temperatures above 25° C in the greenhouse. The most striking symptom is the curving of the terminal bud to form a crook (Plate 20). Later, other buds on the plant become brown, necrotic, and brittle. Such buds may fall off at the slightest touch. Adventitious leaf and floral buds may proliferate excessively (Fig. 62).

The pith of stems and branches may show a brown discoloration, first near the nodes and then throughout the stem (Plate 20). Brown streaks are occasionally observed on petioles and large leaf veins.

Petioles of the youngest trifoliolate leaves are often thickened and shortened and may be curved, distorting the shoot tips. Leaflets are dwarfed and tend to cup or roll, and the blades become more or less rugose and bronzed (Plate 21).

Pods are generally severely underdeveloped or aborted. Those that set before infection often develop dark blotches (Fig. 63 and Plate 22). Such pods generally do not produce viable seeds and drop early. Maturity is delayed in infected plants; they remain green until harvested or killed by frost. TRSV can infect soybeans at any time during the life of the plant, but susceptibility to the virus decreases after bloom.

The virus significantly reduces root and nodule growth. Nodulation is suppressed until plants are about 40 days old. When nodulation does begin, nodule fresh weight and leghemoglobin content are reduced.

In the absence of seedborne inoculum, the disease first appears at the edge of a soybean field and advances inward as the season progresses. The speed of spread depends on the crops and weeds next to the field and probably on insect vector populations. More infection occurs in fields next to pastures, and less occurs in those next to maize fields.

The virus causes systemic infection of susceptible cultivars, moving from infected leaves to the tips of stems and into roots. Movement from roots to leaves is uncommon. Movement is faster at high temperatures and in long photoperiods, and movement from young leaves is greater than that from maturing leaves.

Causal Agent

TRSV is a member of the nepovirus group. The virions are polyhedral particles, 28 to 30 nm in diameter, which can be separated into top-, middle-, and bottom-sedimenting components (53, 91, and 126 S, respectively). The top component is devoid of nucleic acid. The bipartite genome consists of single-stranded RNA molecules with molecular weights of 1.4×10^6 and 2.4×10^6, encapsidated separately in the middle and bottom components, respectively.

The thermal inactivation point of TRSV is between 60 and 65° C. The dilution end point is between 10^{-4} and 10^{-5}, and longevity in vitro is 6–10 days at 25° C and 10 months at 2–4° C.

TRSV is moderately immunogenic. Serological tests have been used for diagnosis and to determine virus concentration in infected plants.

Several strains of TRSV naturally infecting soybeans have been reported. The strains were differentiated by indicator plants.

Host Range

TRSV induces local or systemic symptoms, or both, in a wide range of plants, and some symptomless carriers also are known. The hosts, many of which act as reservoirs of the virus, include *Abutilon theophrastii* Medic., *Amaranthus palmeri* S. Watson, *A. retroflexus* L., *Antirrhinum majus* L., *Apium graveolens* L., *Beta vulgaris* L., *Boerhaavia coccinea* Mill., *B. erecta* L., *Chenopodium album* L., *C. amaranticolor* Coste & Reyn., *C. murale* L., *Citrullus vulgaris* Schrad., *Crotalaria intermedia* Kotschy, *Cucumis melo* L., *C. sativus* L., *Cucurbita pepo* L., *Cyamopsis tetragonoloba* (L.) Taub., *Datura meteloides* DC., *D. stramonium* L., *Gladiolus* spp., *Gomphrena globosa* L., *Helianthus annuus* L., *Hibiscus cannabinus* L., *Iris* spp., *Lactuca sativa* L. var. *capitata*, *Lilium* spp., *Lupinus* spp., *Malvastrum coromandelianum* (L.) Garcke, *Margaranthus solanaceus* Schlect., *Melilotus* spp., *Mentha* spp., *Nicotiana glutinosa* L., *N. repanda* Willd., *N. tabacum* L., *Petunia* ×*hybrida* Hort. Vilm., *P. violacea* Lindl., *Phaseolus lunatus* L., *P. vulgaris* L., *Physalis floridana* Rydb., *Pisum sativum* L., *Polygonum hydropiperoides* Michx., *Portulaca oleracea* L., *Sesamum indicum* L., *Solanum diphyllum* Forsskal, *S. elaeagnifolium* Cav., *S. melongena* L. var. *esculentum* Nees, *Spinacia oleracea* L., *Trifolium pratense* L., *Vigna unguiculata* (L.) Walpers, and *Zinnia elegans* Jacq.

Fig. 62. Bud blight symptoms, caused by tobacco ringspot virus, on a soybean plant in the field. (Courtesy J. B. Sinclair)

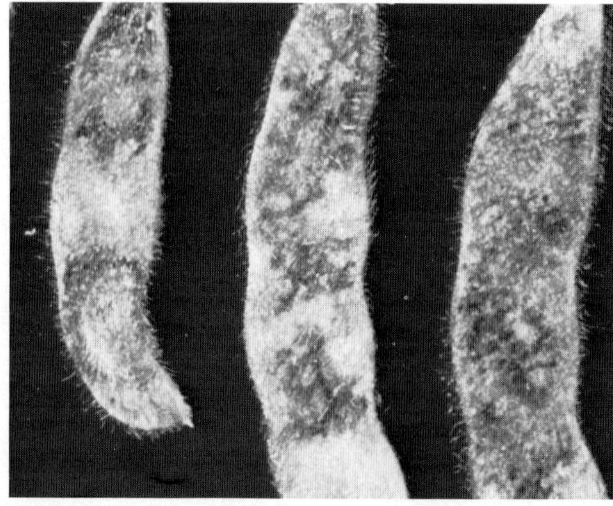

Fig. 63. Soybean pods with symptoms caused by tobacco ringspot virus. (Courtesy M. C. Shurtleff)

Some symptomless carriers are *Ambrosia artemisiifolia* L., *Daucus carota* L., *Erigeron strigosus* Muhl., *Lilium speciosum* Thunb., *Rumex acetosella* L., *Taraxacum officinale* Weber, and *Trifolium repens* L.

Transmission

TRSV is easily sap-transmissible. Circumstantial evidence suggests that the virus may be transmitted by root grafts from infected plants to healthy plants.

No efficient insect vector of TRSV has been discovered. Nymphs (but not adults) of *Thrips tabaci* Lind. transmit it to soybeans at a low level of efficiency. The nymphs appear to retain the virus for at least 14 days after acquisition. A grasshopper (*Melanoplus differentialis* (Thomas)) was capable of low rates of transmission (2–3%) when permitted to take only a single feeding. If the feeding was extended to 30 sec, no transmission occurred. Thus, transmission by grasshoppers in nature is questionable, because of their naturally voracious eating habits.

The dagger nematode *Xiphinema americanum* Cobb is also a vector of TRSV, but its efficiency in transmitting the virus to soybeans is low. Even when it does occur, nematode transmission of TRSV to roots of soybean plants may be of no significance, because the infection generally remains confined to the roots.

Seed transmission is the most important mode of long-range dissemination and carry-over from season to season. Systemically infected plants often produce infected seeds, which give rise to diseased seedlings. The extent of seed transmission depends on when infection takes place; plants infected before bloom produce few or no seeds. Because all infected seeds come from plants infected before bloom, the maximum possible amount of viruliferous seeds is extremely small, barely enough to perpetuate the virus in a few widely scattered plants. The proportion of diseased seeds is greater when the time between infection and flowering is long. The location of pods on the plant and of seeds in the pod does not have a significant effect on the frequency of seed infection. Less seed infection occurs during prolonged cool periods. The virus, which remains viable in seeds for at least 5 years, is carried in the periplasm and embryo but not the seed coat.

Control

A few soybean cultivars have been noted to have resistance to a few strains of TRSV. Furthermore, one genotype (PI 407287) of wild soybean (*Glycine soja* Sieb. & Zucc.) is resistant to the virus.

Virus-free soybean seeds should be used in commercial fields, and it may be desirable to avoid fields with dagger nematodes or treat them with an appropriate nematicide.

Since the speed of spread of TRSV depends on the crops and weeds next to soybean fields and probably also on insect vector populations, it may be desirable to locate soybean fields next to maize fields rather than pastures.

Selected References

Allington, W. B. 1946. Bud blight of soybean caused by the tobacco ring-spot virus. Phytopathology 36:319-322

Athow, K. L., and Laviolette, F. A. 1961. The relation of seed-transmitted tobacco ringspot virus to soybean yield. Phytopathology 51:341-342.

Crittenden, H. W., Hastings, K. M., and Moore, G. M. 1966. Soybean losses caused by tobacco ringspot virus. Plant Dis. Rep. 50:910-913.

Gergerich, R., Asher, J. H., Jr., and Ramsdel, D. C. 1983. A comparison of some serological and biological properties of seven isolates of tobacco ringspot virus. Phytopathol. Z. 107:289-300.

McGuire, J. M. 1964. Efficiency of *Xiphinema americanum* as a vector of tobacco ringspot virus. Phytopathology 54:799-801.

Messieha, M. 1969. Transmission of tobacco ringspot virus by thrips. Phytopathology 59:943-945.

Orellana, R. G. 1981. Resistance to bud blight in introductions from the germ plasm of wild soybean. Plant Dis. 65:594-595.

Tu, J. C. 1986. Strains of tobacco ringspot virus isolated from soybean (*Glycine max*) in southwestern Ontario (Canada). Can. J. Plant Sci. 66:491-498.

(Prepared by J. W. Demski and C. W. Kuhn)

Tobacco Streak Virus

Tobacco streak virus (TSV) causes Brazilian bud blight of soybeans. It was first reported in Brazil in 1950. The disease is common and destructive in soybeans in southeastern Brazil and also occurs in Iowa and Oklahoma. It is indistinguishable from the bud blight caused by tobacco ringspot virus (see Tobacco Ringspot Virus).

Symptoms

TSV symptoms are usually not seen on young plants. Irregular, yellow spots form later on leaves, followed by systemic symptoms. Infected plants tend to recover and then develop supernumerary axillary branches, which are stunted and produce dwarfed leaves (Plate 23). Mosaic symptoms and necrotic streaks may develop at nodes. Necrotic blotches appear on pods. Early-infected plants produce fewer pods and seeds. Infection at any age delays seed maturation.

TSV moves rapidly in soybeans from roots to aerial parts and from leaves to shoots and roots. It can be recovered from all parts of the plant except pollen. Virus concentration increases in young developing leaves and declines as they mature.

Causal Agent

TSV, a member of the ilarvirus group, has isometric particles ranging from 25 to 30 nm in diameter. The particles separate into three or four sedimenting components, with sedimentation coefficients ranging from 78 to 114 S. Particles isolated from infected plants contain four molecules of positive-sense, single-stranded RNA ranging in molecular weight from 0.3×10^6 to 1.04×10^6. The three largest RNAs are genomic, but they require either coat protein or the smallest (subgenomic) RNA for infectivity.

Many strains are known, and cross-protection between strains is common. The thermal inactivation point of TSV at pH 7 is between 55 and 60° C in phosphate buffer and in sodium sulfite solution. The dilution end point is about 10^{-1} in sap and 10^{-5} to 10^{-6} if the virus is extracted with phosphate buffer and sodium sulfite. A dilution end point of 1:640 was reported in phosphate buffer. The virus is inactivated within 20 min in crude sap and within 1 hr if extracted with water; infectivity is retained for up to 9 hr if it is extracted with phosphate buffer and sodium sulfite.

Host Range

TSV has a wide host range, including peanut (*Arachis hypogaea* L.), asparagus (*Asparagus officinalis* L.), papaya (*Carica papaya* L.), dahlia (*Dahlia variabilis* (Willd.) Desf.), cotton (*Gossypium hirsutum* L.), tomato (*Lycopersicon esculentum* Mill.), tobacco (*Nicotiana* spp.), pea (*Pisum sativum* L.), rose (*Rosa* spp.), raspberry (*Rubus* spp.), potato (*Solanum tuberosum* L.), and grape (*Vitis* spp.).

Transmission

TSV is sap-transmissible but is readily inactivated if extracted in water. Infectivity can be preserved by extracting the virus in the presence of sodium sulfite or other reducing agents.

The virus is seed-transmitted. The extent of seed transmission depends on the virulence of the virus strain, susceptibility of the cultivar, and earliness of infection. In early-infected susceptible cultivars, seed transmission up to 30% has been recorded. The virus is found in the embryo and seed coat of mature seeds.

Indirect evidence suggests transmission by a thrips, *Caliothrips phaseoli* (Hood).

Control

Resistance in soybeans to TSV has not been reported. Therefore, control measures should emphasize protection from introducing TSV-infected seeds into soybean fields. In areas where the virus causes significant problems in soybean performance, the production of virus-free seeds for planting should be very useful.

Selected References

Costa, A. S., and Carvalho, A. M. B. 1961. Studies on Brazilian tobacco streak. Phytopathol. Z. 42:113-138.
Fagbenle, H. H., and Ford, R. E. 1970. Tobacco streak virus isolated from soybeans, *Glycine max*. Phytopathology 60:814-820.
Ghanekar, A. M., and Schwenk, F. W. 1980. Comparison of tobacco streak virus isolates from soybean and tobacco. Phytopathol. Z. 97:148-155.
Sherwood, J. L., and Jackson, K. E. 1985. Tobacco streak virus in soybean in Oklahoma. Plant Dis. 69:727.

(Prepared by J. W. Demski and C. W. Kuhn)

Bean Yellow Mosaic Virus

Bean yellow mosaic virus (BYMV) causes a mosaic and yellowing disease of soybeans (Fig. 64). Mixed infections of BYMV and soybean mosaic virus are common in soybeans. Reduced oil content and increased protein content have been recorded in seeds from BYMV-infected plants.

Symptoms

The initial symptoms are essentially the same as those produced by soybean mosaic virus at later stages (see Soybean Mosaic Virus). A conspicuous yellow mottling of the leaves is characteristic (Plates 24 and 25). The yellow areas are either scattered or produced in indefinite bands along the major veins. Rusty, necrotic spots appear in the yellow areas as the leaves mature. Some strains of BYMV produce severe mottling and crinkling of leaves (Plates 24 and 25).

Causal Agent

BYMV, a member of the potyvirus group, has flexuous-rod particles, about 750×12 nm. Its dilution end point is between

Fig. 64. Symptoms caused by bean yellow mosaic virus on a trifoliolate soybean leaf. (Courtesy J. B. Sinclair)

10^{-3} and 10^{-4}. The thermal inactivation point varies from 50 to 62°C, depending on the viral strain. Similarly, longevity in vitro at room temperature varies from 1 to 4 days.

Transmission

BYMV has a wide host range and is easily sap-transmissible. More than 20 aphid species can transmit the virus in a nonpersistent manner. Common aphid vectors include *Acyrthosiphon pisum* (Harris), *Aphis fabae* Scopoli, *Macrosiphum euphorbiae* (Thomas), and *Myzus persicae* (Sulzer). Seed transmission in soybeans has not been reported.

Selected References

Afanasiev, M. M., and Morris, H. E. 1952. Bean virus 2 (yellow) on Great Northern bean in Montana. Phytopathology 42:101-104.
Conover, R. A. 1948. Studies of two viruses causing mosaic diseases of soybean. Phytopathology 38:724-735.
Quantz, L. 1961. Untersuchungen über das gewöhnlich bohnenmosaik-virus und das sojamosaikvirus. Phytopathol. Z. 43:79-101.

(Prepared by J. W. Demski and C. W. Kuhn)

Black Gram Mottle Virus

Black gram mottle virus (BGMV) causes a mild mottle disease of soybeans in Thailand. The virus is mechanically transmissible and has isometric particles about 28 nm in diameter. It has one nucleoprotein component, with a sedimentation coefficient of 122 S and a density of $1.364\,\mathrm{g/cm^3}$. It has a thermal inactivation point between 85 and 90°C, a dilution end point between 10^{-9} and 10^{-10}, and longevity in vitro of 6 to 9 weeks at 20°C.

The virus can infect plants in at least six families. Most systemic infections occur in the Fabaceae. Susceptible soybean genotypes react with a systemic mild mottle, and resistant ones with necrotic local lesions.

BGMV is transmitted by the beetles *Monolepta signata* Olivier, *Cerotoma trifurcata* (Forster), and *Epilachna varivestis* Mulsant. However, it is serologically unrelated to three other beetle-transmitted viruses: bean mild mosaic virus, cowpea mottle virus, and southern bean mosaic virus.

Selected References

Honda, Y., Iwaki, M., Thongmeearkom, P., Deema, N., and Srithongchai, W. 1982. Blackgram mottle virus occurring on mungbean and soybean in Thailand. JARQ 16:72-77.
Scott, H. A., and Phatak, H. C. 1979. Properties of blackgram mottle virus. Phytopathology 69:346-348.

(Prepared by J. W. Demski and C. W. Kuhn)

Cowpea Mild Mottle Virus

Cowpea mild mottle virus (CMMV) has been isolated from soybeans in Indonesia, Malaysia, and Thailand and appears to be widely distributed in tropical countries. Symptoms in soybeans vary according to the cultivar. Veinclearing and leaf rolling are characteristic in the cultivar Shirotsurunoko; mosaic and vein necrosis develop in the cultivar Toyosuza.

CMMV is a carlavirus with rod-shaped particles 650–700 nm long. It has a thermal inactivation point near 70°C and a dilution end point between 10^{-5} and 10^{-6}. It can remain infectious for up to 21 days at 20°C. In ultrathin sections of soybean leaves, featherlike and bundle inclusion bodies were observed in the cytoplasm.

The virus is mechanically transmitted and can be transmitted by the whitefly *Bemisia tabaci* (Gennadius) but not by aphids. It has a moderately wide host range in five plant families.

Selected References

Iwaki, M., Thongmeearkom, P., Honda, Y., Prommin, M., Deema, N., Hibi, T., Iezuka, N., Ong, C. A., and Saleh, N. 1986. Cowpea mild mottle virus occurring on soybean and peanut in southeast Asian countries. Pages 106-120 in: Virus Diseases of Rice and Legumes in the Tropics. Tech. Bull. 21. Tropical Agriculture Research Center, Yatabe, Tsukuba, Ibaraki, Japan.

Iwaki, M., Thongmeearkom, P., Prommin, M., Honda, Y., and Hibi, T. 1982. Whitefly transmission and some properties of cowpea mild mottle virus on soybean in Thailand. Plant Dis. 66:365-368.

(Prepared by J. W. Demski and C. W. Kuhn)

Cowpea Severe Mosaic Virus

Cowpea severe mosaic virus (CSMV) has been isolated from soybeans from widely separated geographic areas, such as Brazil, Puerto Rico, Trinidad, and the United States. It causes severe mosaic and stunting. Soybean plants inoculated 14 days after planting often die prematurely.

The causal agent belongs to the comovirus group. Its particles are isometric, with a diameter of about 25 nm. Isolates from different geographic areas are serologically related to each other (although not necessarily homologous) and often weakly to other comoviruses. The stability of CSMV in sap is somewhat variable, with the thermal inactivation point ranging from 65 to 70°C, the dilution end point from 10^{-4} to 10^{-5}, and longevity in vitro from 1 to 5 days.

The virus is primarily restricted to leguminous plants. *Vigna unguiculata* (L.) Walpers is a diagnostic and propagation host; *Chenopodium amaranticolor* Coste & Reyn. is a local lesion host.

CSMV can be mechanically transmitted and beetle-transmitted. Numerous leaf-feeding beetles, primarily in the Chrysomelidae, are the main vectors. The virus is seed-transmitted in cowpea but not in soybeans.

Selected References

Anjos, J. R. N., and Lin, M. T. 1984. Bud blight of soybeans caused by cowpea severe mosaic virus in central Brazil. Plant Dis. 68:405-407.

McLaughlin, M. R., Thongmeearkom, P., Goodman, R. M., Milbrath, G. M., Ries, S. M., and Royse, D. J. 1978. Isolation and beetle transmission of cowpea mosaic virus (severe subgroup) from *Desmodium canescens* and soybeans in Illinois. Plant Dis. Rep. 62:1069-1073.

(Prepared by J. W. Demski and C. W. Kuhn)

Indonesian Soybean Dwarf Virus

Indonesian soybean dwarf virus (ISDV) has been isolated from soybeans in Indonesia and Thailand. The virus induces dwarfing, with shortened leaf petioles and internodes, and infected plants are often dark green. Upper leaves are small and often crinkled and curled. Lower leaves show rugosity and are often brittle.

ISDV has isometric particles 25–30 nm in diameter. Aggregates of particles occur in vacuoles of phloem parenchyma cells. The virus is not mechanically transmissible but can be transmitted by *Aphis glycines* Matsumura in a persistent manner. Particle morphology, mode of transmission, and tissue association suggest that ISDV is a member of the luteovirus group. However, no serological relationship has been noted between ISDV and soybean dwarf virus isolated in Japan.

Selected References

Honda, Y., Iwaki, M., Goto, T., Thongmeearkom, P., Deema, N., Surin, P., Sarindu, N., and Tochebaren, H. 1986. The occurrence of Indonesian soybean dwarf virus on soybean in Thailand. Pages 126-131 in: Virus Diseases of Rice and Legumes in the Tropics. Tech. Bull. 21. Tropical Agriculture Research Center, Yatabe, Tsukuba, Ibaraki, Japan.

Iwaki, M., Roechan, M., Hibino, H., Tochihara, H., and Tantera, D. M. 1980. A persistent aphidborne virus of soybean, Indonesian soybean dwarf virus. Plant Dis. 64:1027-1030.

(Prepared by J. W. Demski and C. W. Kuhn)

Mung Bean Yellow Mosaic Virus

Mung bean yellow mosaic virus (MYMV) causes considerable damage to mung bean in India and Thailand and has been isolated from naturally infected soybeans. The initial symptom in systemically infected soybean leaves is vein yellowing along veinlets; severe yellow mosaic develops in older infections.

MYMV has geminate particles, 18×30 nm, a thermal inactivation point between 40 and 50°C, and a dilution end point between 10^{-2} and 10^{-3}. Loose aggregates of particles can be observed in the nucleus of phloem cells.

MYMV can be transmitted mechanically and by the whitefly *Bemisia tabaci* (Gennadius). Its host range appears to be limited to the Fabaceae.

Selected References

Honda, Y., Iwaki, M., Thongmeearkom, P., Kiratiya-Angul, K., Kiratiya-Angul, S., Srithongchai, W., Prommin, M., Kitipakorn, K., Sarindu, N., Deema, N., Syamananda, R., Hibi, T., and Saito, Y. 1986. Mungbean yellow mosaic virus isolated from mungbean in Thailand. Pages 189-202 in: Virus Diseases of Rice and Legumes in the Tropics. Tech. Bull. 21. Tropical Agriculture Research Center, Yatabe, Tsukuba, Ibaraki, Japan.

Singh, B. B., and Malick, A. S. 1978. Inheritance of resistance to yellow mosaic in soybean. Indian J. Genet. Plant Breed. 38:258-261.

(Prepared by J. W. Demski and C. W. Kuhn)

Peanut Stripe Virus

A virus producing mild mottle (VPMM) in peanut and also naturally infecting soybeans was reported in Hubei Province, China. The virus reduced soybean yields by as much as 53% in greenhouse trials. J. W. Demski and co-workers reported a virus named peanut stripe virus (PSTV) naturally infecting peanut in the United States. PSTV infects soybeans and is considered to be the same virus as VPMM.

Symptoms

Symptoms of PSTV vary with the soybean cultivar. Only a mild systemic mottle develops in the cultivar Lawrence, systemic mosaic in Elgen, chlorotic local lesions followed by systemic mottle in Coc Chun, and necrotic local lesions followed by systemic necrosis in Tulumayo-2.

Causal Agent

PSTV is a potyvirus with flexuous-rod particles about 750 nm long. It has a thermal inactivation point near 60°C and a dilution end point between 10^{-3} and 10^{-4}. The virus remains infective in crude sap for 3 days at 20°C. PSTV is serologically related to blackeye cowpea mosaic virus, clover yellow vein virus, and soybean mosaic virus.

Host Range and Transmission

PSTV infects numerous legumes, such as soybeans, peanut, blue and white lupines, and clovers (crimson, subterranean, and arrowleaf). Its host range is not restricted to legumes. *Chenopodium amaranticolor* Coste & Reyn. is a local lesion host.

The virus is seed-transmitted in peanut but not soybeans. It is easily transmitted mechanically and by aphids.

Control

Planting soybeans over 100 m from infected peanut reduces the chance of infection.

Selected References

Demski, J. W., Reddy, D. V. R., Sowell, G., Jr., and Bays, D. 1984. Peanut stripe virus—A new seed borne potyvirus from China infecting groundnut (*Arachis hypogaea*). Ann. Appl. Biol. 105:495-501.

Xu, Z., Yu, Z., Liu, J., and Barnett, O. W. 1983. A virus causing peanut mild mottle in Hubei Province, China. Plant Dis. 67:1029-1032.

(Prepared by J. W. Demski and C. W. Kuhn)

Soybean Chlorotic Mottle Virus

Soybean chlorotic mottle virus is a caulimovirus isolated in Japan from naturally infected soybeans with mosaic and stunting symptoms. In the greenhouse, soybean cultivars showed veinclearing and chlorosis followed by mottle; leaf size was reduced, and infected plants were stunted, with shortened internodes.

Purified virus preparations contained spherical particles about 50 nm in diameter, which contained double-stranded DNA. The virus was found to be not serologically related to cauliflower mosaic virus. Inclusion bodies in soybean and bean tissues were frequently ovoid or elliptical and consisted of an amorphous, vacuolated, electron-dense matrix.

The host range was restricted to four plant species—*Glycine max* (L.) Merrill, *Lablab purpureus* (L.) Sweet, *Phaseolus vulgaris* L., and *Vigna unguiculata* (L.) Walpers—out of 42 species within 12 families tested. No seed transmission was detected. The causal agent is mechanically transmissible but could not be transmitted by five species of aphids.

Selected Reference

Iwaki, M., Isogawa, Y., Tsuzuki, H., and Honda, Y. 1984. Soybean chlorotic mottle, a new caulimovirus on soybean. Plant Dis. 68:1009-1011.

(Prepared by J. W. Demski and C. W. Kuhn)

Soybean Crinkle Leaf Virus

A virus causing crinkle leaf disease was detected in most soybean-growing areas of Thailand. Leaves showed crinkling, curling, cupping, and twisting and had veinal enations on the undersurface.

Purified preparations from diseased plants contained geminate particles, 8×30 nm.

Soybean crinkle leaf virus was transmitted by whiteflies and grafting but not mechanically or by aphids. Whitefly transmission caused systemic symptoms in soybeans and in *Cassia tora* L., *Datura stramonium* L., *Lycopersicon esculentum* Mill., *Nicotiana clevelandii* A. Gray, *N. debneyi* Domin, *N. glutinosa* L., *N. tabacum* L., *Petunia* ×*hybrida* Hort. Vilm., *Phaseolus vulgaris* L., and *Zinnia elegans* Jacq.

Selected Reference

Iwaki, M., Thongmeearkom, P., Honda, Y., Sarindu, N., Deema, N., and Surin, P. 1986. Soybean crinkle leaf disease occurring on soybean in Thailand. Pages 132-143 in: Virus Diseases of Rice and Legumes in the Tropics. Tech. Bull. 21. Tropical Agriculture Research Center, Yatabe, Tsukuba, Ibaraki, Japan.

(Prepared by J. W. Demski and C. W. Kuhn)

Soybean Yellow Vein Virus

Soybean yellow vein virus (SYVV), causing a yellowing of veins on the upper leaves of soybeans in the field and in the greenhouse, has been isolated in central Thailand. The virus is rod-shaped, 500–550 nm long, and can be mechanically transmitted. It may be a member of the tobamovirus group. No insect vectors have been found, and the host range is limited to soybeans, in which the virus produces systemic infections, and *Chenopodium* sp., in which it produces local lesions.

Selected Reference

Senbokee, T., Thongmeearkom, P., Kitipakorn, K., Kiratiya-Angul, S., Srithongchai, W., and Deema, N. 1986. Soybean yellow-vein virus, a new virus occurring on soybean in Thailand. Pages 121-125 in: Virus Diseases of Rice and Legumes in the Tropics. Tech. Bull. 21. Tropical Agriculture Research Center, Yatabe, Tsukuba, Ibaraki, Japan.

(Prepared by J. W. Demski and C. W. Kuhn)

Tobacco Mosaic Virus

A new strain of tobacco mosaic virus (TMV), a member of the tobamovirus group, has been isolated from naturally infected soybeans in Yugoslavia. The soybean strain (TMV-S) induced veinclearing and mild chlorotic mosaic symptoms.

The causal agent has a thermal inactivation point of 90–95° C and a dilution end point between 10^{-6} and 10^{-7}. Its longevity in vitro is approximately 13 days. In immunodiffusion tests, TMV-S reacted with antisera to itself and to the common strain of TMV but did not react with antiserum to the bean strain of the virus. It caused the formation of hexagonal crystals in tobacco leaf trichomes, which are characteristic of various TMV strains.

TMV-S causes symptoms similar to those caused by the common strain of TMV in *Nicotiana glutinosa* L., *N. tabacum* L. cv. Xanthi, *N. edwardsoniana* Christie & D. W. Hall, *Datura stramonium* L., and *Gomphrena globosa* L. It infects *Pisum sativum* L. and *Trifolium pratense* L. but does not cause infection in *Phaseolus vulgaris* L. cvs. Topcrop, Tendercrop, Pinto, or Bountiful, *Vigna unguiculata* (L.) Walpers cv. Blackeye, or *Lycopersicon esculentum* L. cvs. Rutgers or Marglobe. TMV-S is mechanically transmitted but not transmitted through soybean seeds.

Selected Reference

Taraku, N., and Tolin, S. A. 1986. Properties of a strain of tobacco mosaic virus from soybean in Yugoslavia. (In Yugoslavian) Zast. Bilja 37:175-185.

(Prepared by J. W. Demski and C. W. Kuhn)

The following viruses have been reported to naturally infect soybeans, but very little information is available about their economic importance: Abutilon mosaic (Plate 26), beet western yellows, horse gram yellow mosaic, milk vetch dwarf, passion fruit woodiness, soybean mild mosaic, soybean stunt, and tomato spotted wilt viruses.

Boswell, K. F., and Gibbs, A. J. 1983. Virus Identification Data Exchanges. Canberra Publishing and Printing, Canberra, Australia. 139 pp.

(Prepared by J. W. Demski and C. W. Kuhn)

Nematode Diseases

Nematodes, also called eelworms or nemas, are unsegmented roundworms that inhabit fresh and salt water, decaying organic matter, soil, plants, and animals, including humans. They are probably the most numerous multicellular animals on earth. Vast numbers of various soil-inhabiting forms can be found in every field, regardless of the crop. A 500-cm^3 (1-pint) soil sample may contain 20,000 or more nematodes. Nematologists estimate that 3 billion nematodes live in 0.4 ha (1 acre) of typical soil, mostly within the top 15–20 cm.

Most of the more than 15,000 described species are microscopic, transparent, vermiform, and mobile. Several rather sluggish ectoparasites, such as *Criconemoides* and *Macroposthonia* spp., and females of *Heterodera* and *Meloidogyne* spp., whose bodies become swollen and saclike, are exceptions. Nematodes are identified by anatomic features such as the shape and size of the adult female, the shape and size of the stylet and tail, the shape of the esophagus and reproductive organs, and the pattern of the cuticle.

Active movement of nematodes in soil is limited. They rarely move more than 75 cm per year. However, they move passively in water, soil, and infected plant parts and are disseminated by the wind, many types of animals, tools, and vehicles and machinery carrying infested soil.

Normally, mixtures of parasitic and nonparasitic nematodes occur in, on, or about soybean roots. Flotation, sieving, and centrifugation procedures are used to extract nematodes from soil; incubation techniques, such as incubation in mist chambers, are used to extract them from roots. The pathogenicity of plant-parasitic species is determined by their feeding habits, the relationship of symptoms above and below ground to the numbers present, and the frequency with which a particular nematode is associated with a particular symptom. Usually the rhizosphere of soybean plants must be inoculated with individual species to demonstrate the virulence of the nematode population.

The optimal temperature for activity of most species of nematodes is between 16 and 32°C. However, it varies greatly with the species, stage of nematode development, and growth of the host plant. Most plant-parasitic nematodes become inactive at temperatures below 10°C or above 35°C.

More than 100 species of plant-parasitic nematodes have been reported to feed on or be associated in some way with the roots of soybean plants, but only a few are of economic importance. The current classification of plant-parasitic nematodes associated with soybeans is presented in Table 9. Feeding by the parasites decreases the efficiency of the root system in the absorption and transport of water and nutrients to the shoot. Disruption of normal root functions suppresses growth and is often accompanied by chlorosis or other nutrient deficiency symptoms; ultimately, yields are lowered. The degree of damage is related to the cropping history of the field, tillage practices, the population of nematodes in the soil at planting time, and environmental conditions after the plants emerge. Usually soybean plants are most susceptible to nematode injury when young and become more tolerant with age. Soil hardpans, mineral deficiencies, drought stress, herbicide injury, and abnormally high temperatures weaken plants and render them more susceptible to nematodes.

Virtually all soybean fields contain more than one species of plant-parasitic nematode. The kinds and numbers of nematodes in the soil, relative to the age and condition of the plants, and environmental stresses on the plants are important considerations in determining whether nematodes cause economic injury to the crop. Damage often reflects the additive effects of several species feeding on a root system, as well as their interactions with other root-infecting pathogens.

Generalized Life Cycle

The life cycles of most plant-parasitic nematodes, though varying in details, are relatively simple (Fig. 65). Juveniles (larvae) hatch from eggs deposited by females in the soil or in root tissue. Juveniles usually resemble adults in structure and appearance; they develop through four stages, each completed by a molt, reaching maturity after the final molt. Mature nematodes reproduce either sexually or parthenogenetically and lay eggs to start the next generation. Under optimal conditions, most plant-parasitic species complete their life cycles in 3 or 4 weeks. Three or four generations may be produced in one growing season. A few species, such as *Xiphinema americanum* Cobb, apparently produce only one generation per year and may require 9 months or longer to complete a cycle.

Feeding Habits

All nematodes that attack soybeans are obligate parasites that must feed on living plants to complete their life cycle. Almost all feed on plant cells by puncturing cell walls with a hollow stylet (which resembles a minute hypodermic needle), injecting enzymes into the cells, and then ingesting the partially digested contents.

Plant-parasitic nematodes are divided into two major groups, on the basis of their parasitic behavior. Ectoparasitic nematodes, such as *Belonolaimus, Paratrichodorus*, and *Xiphinema* spp., spend their lives outside their hosts, browsing on roots and feeding by inserting their stylets to varying depths into root tissues. Endoparasites feed internally, within plants. Migratory endoparasites, such as *Pratylenchus* spp., remain mobile. Sedentary endoparasites, such as *Meloidogyne* spp., become immobile and assume a sedentary parasitic habit. Some sedentary forms, such as *Heterodera* and *Rotylenchulus* spp., are semiendoparasites, spending most of their lives partially embedded in roots. Endoparasitic nematodes such as *Heterodera* and *Meloidogyne* spp. are recognized more easily than the ectoparasitic forms.

Nematodes damage soybean roots directly or by interacting with other pathogens. They injure or kill root cells mechanically or chemically (apparently by secreting or injecting enzymes). Such damage lessens the efficiency of the roots in absorbing and transporting water and nutrients. In many cases, root growth ceases. Nematodes also interact with bacterial or fungal pathogens in disease complexes, which often damage roots more severely than either the nematode or the other pathogen alone.

Symptoms of Nematode Injury

The feeding of nematodes on soybean roots reduces root growth and plant vigor, lowers the natural resistance of plants to other pathogens, and induces root lesions, galls, or other deformations. The damage commonly ranges from barely detectable injury to obvious yield reduction. It may be restricted to part of a field or extend over an entire field. Striking symptoms may not be present if nematode populations are too low to cause detectable damage. In heavily infested fields, plant growth appears uneven, usually with distinct areas of stunted, chlorotic plants (Plate 27). The type and severity of symptoms vary with the nematode species and population density, the age and condition of the plants, and soil and climatic conditions.

In more severe cases, primary symptoms on soybeans may include combinations of the following abnormalities:

Stunting or suppressed growth. Plants are uneven in height, usually in irregular spots; the field looks "ragged."

Chlorosis or other discoloration. Yellowing resembles symptoms of nitrogen, iron, magnesium, manganese, or zinc deficiency; purple coloration resembles that caused by phosphorus deficiency.

Wilting. Wilting is enhanced by wind and bright sunlight. Plants are often slow to recover at night.

Root lesions or dark, discolored areas in roots. As nematodes continue feeding, lesions grow from pinpoint spots to large necrotic areas. Root-rot fungi often invade tissues damaged by nematodes, accelerating lesion enlargement.

Root swellings. Swellings vary from indistinct enlargements, primarily of the root tips, to knotlike galls throughout the root system, which are often accompanied by excessive production of branch roots (a condition known as hairy root) above the swelling.

Injured or devitalized root tips. Nematode feeding at or near root tips arrests root growth. Tissues may retain normal coloration or turn brown and die. Devitalization frequently causes excessive production of roots near the soil surface and stubby or coarse roots or both on the rest of the root system (Fig. 65). A stubby root system is composed of numerous short, stubby branches, often arranged in clusters. A coarse root system has few or no branch roots or feeder rootlets.

Reduced and discolored root system. An overall reduction may occur in the size of the root system, including nodulation, accompanied by light to dark browning or blackening.

Yield reductions. Weakened plants produce smaller and fewer pods and less forage or green manure. Heavily damaged plants stressed by other factors may produce empty pods or none at all.

Because symptoms of nematode damage often mimic those induced by low or unbalanced fertility, poor drainage, drought, soil insects, root-rot fungi, or herbicides, diagnosis of nematode diseases should never be based solely on plant symptoms. Late

---

Table 9. Classification of Plant-Parasitic Nematodes Associated with Soybeans

Phylum: Nematoda	Order: Tylenchida (*continued*)
Order: Dorylaimida	Suborder: Tylenchina (*continued*)
Suborder: Dorylaimina	Superfamily: Tylenchoidea (*continued*)
Superfamily: Dorylaimoidea	Family: Hoplolaimidae
Family: Longidoridae	Genus: *Hoplolaimus* (lance nematodes)
Genus: *Xiphinema* (dagger nematodes)	Species: *H. columbus*
Species: *X. americanum*	*H. galeatus*
X. diversicaudatum	*H. magnistyles*
Suborder: Diphtherophorina	Genus: *Rotylenchus* (spiral nematodes)
Superfamily: Trichodoroidea	Species: *Rotylenchus* sp.
Family: Trichodoridae	Genus: *Helicotylenchus* (spiral nematodes)
Genus: *Paratrichodorus* (stubby-root nematodes)	Species: *H. delhiensis*
Species: *P. christiei*	*H. dihystera*
P. porosus	*H. paragirus*
Order: Tylenchida	*H. pseudorobustus*
Suborder: Tylenchina	Genus: *Scutellonema* (spiral nematodes)
Superfamily: Tylenchoidea	Species: *S. brachyurum*
Family: Anguinidae	*S. cavenessi*
Genus: *Ditylenchus* (stem and bulb nematodes)	Genus: *Rotylenchulus* (reniform nematodes)
Species: *D. dipsaci*	Species: *R. reniformis*
Family: Belonolaimidae	Family: Heteroderidae
Genus: *Belonolaimus* (sting nematodes)	Genus: *Heterodera* (cyst nematodes)
Species: *B. gracilis*	Species: *H. glycines*
B. longicaudatus	Genus: *Meloidogyne* (root-knot nematodes)
Genus: *Tylenchorhynchus* (stunt nematodes)	Species: *M. arenaria*
Species: *T. agri*	*M. hapla*
T. capitatus	*M. incognita*
T. claytoni	*M. javanica*
T. martini	Superfamily: Criconematoidea
T. nudus	Family: Criconematidae
T. vulgaris	Genus: *Criconemella* (ring nematodes)
Family: Pratylenchidae	Species: *C. ornata*
Genus: *Pratylenchus* (lesion nematodes)	*C. simile*
Species: *P. agilis*	Family: Tylenchulidae
P. alleni	Genus: *Paratylenchus*
P. brachyurus	Species: *P. projectus*
P. coffeae	
P. crenatus	
P. hexincisus	
P. neglectus	
P. pandas	
P. penetrans	
P. safaensis	
P. scribneri	
P. vulnus	
P. zeae	

in the season, nematode damage is especially difficult to distinguish from injury caused by other pathogens and pests. Accurate diagnosis of nematode problems requires analysis by a nematologist, including proper sample collection and handling of soil and roots.

Selected References

Dropkin, V. H. 1980. Introduction to Plant Nematology. John Wiley & Sons, New York. 293 pp.

Edwards, D. I. 1976. Nematodes and nematode-fungus interactions on soybeans. Pages 629-633 in: World Soybean Research. L. W. Hill, ed. Interstate Printers & Publishers, Danville, IL.

Edwards, D. I., and Epps, J. M. 1975. Annotated bibliography of nematodes of soybeans, 1969–1973. U.S. Dep. Agric., Agric. Res. Serv., Publ. NC 24.

Epps, J. M., Edwards, D. I., Good, J. M., and Rebois, R. V. 1973. Annotated bibliography of nematodes of soybeans, 1882–1968. U.S. Dep. Agric., Agric. Res. Serv., Publ. S-8.

Good, J. M. 1973. Nematodes. Pages 537-543 in: Soybeans: Improvement, Production, and Uses. B. E. Caldwell, ed. Agron. Publ. 16. American Society of Agronomy, Madison, WI.

Mai, W. F., and Lyon, H. H. 1975. Pictorial Key to Genera of Plant-Parasitic Nematodes, 4th ed. Comstock and Cornell University Press, Ithaca, NY.

Norton, D. C. 1978. Ecology of Plant-Parasitic Nematodes. John Wiley & Sons, New York.

Schmitt, D. P., and Noel, G. R. 1984. Nematode parasites of soybeans. Pages 13-59 in: Plant and Insect Nematodes. W. R. Nickle, ed. Marcel Dekker, New York.

Tisselli, O., Sinclair, J. B., and Hymowitz, T. 1980. Sources of Resistance to Selected Fungal, Bacterial, Viral and Nematode Diseases of Soybeans. INTSOY Ser. 18. College of Agriculture, University of Illinois at Urbana-Champaign. 134 pp.

Webster, J. M., ed. 1972. Economic Nematology. Academic Press, New York. 563 pp.

Zuckerman, B. M., Mai, W. F., and Harrison, M. B., eds. 1985. Plant Nematology Laboratory Manual. University of Massachusetts, Agricultural Experiment Station, Amherst. 212 pp.

Zuckerman, B. M., Mai, W. F., and Rohde, R. A., eds. 1971. Plant Parasitic Nematodes, Vols. I and II. Academic Press, New York.

(Prepared by G. R. Noel and D. I. Edwards)

Soybean Cyst Nematode

Circular patches of stunted, yellow soybeans, observed in Japan as early as 1881, were ascribed to a disease called moon night or yellow dwarf disease. The disease was not reported until 1915. The causal agent was at first thought to be the sugar beet cyst nematode, *Heterodera schachtii* Schmidt. In 1952, however, it was described and named the soybean cyst nematode, *H. glycines* Ichinohe. It occurs in Canada, the People's Republic of China, Colombia, Indonesia, Japan, Korea, the Soviet Union, and 26 states in the United States.

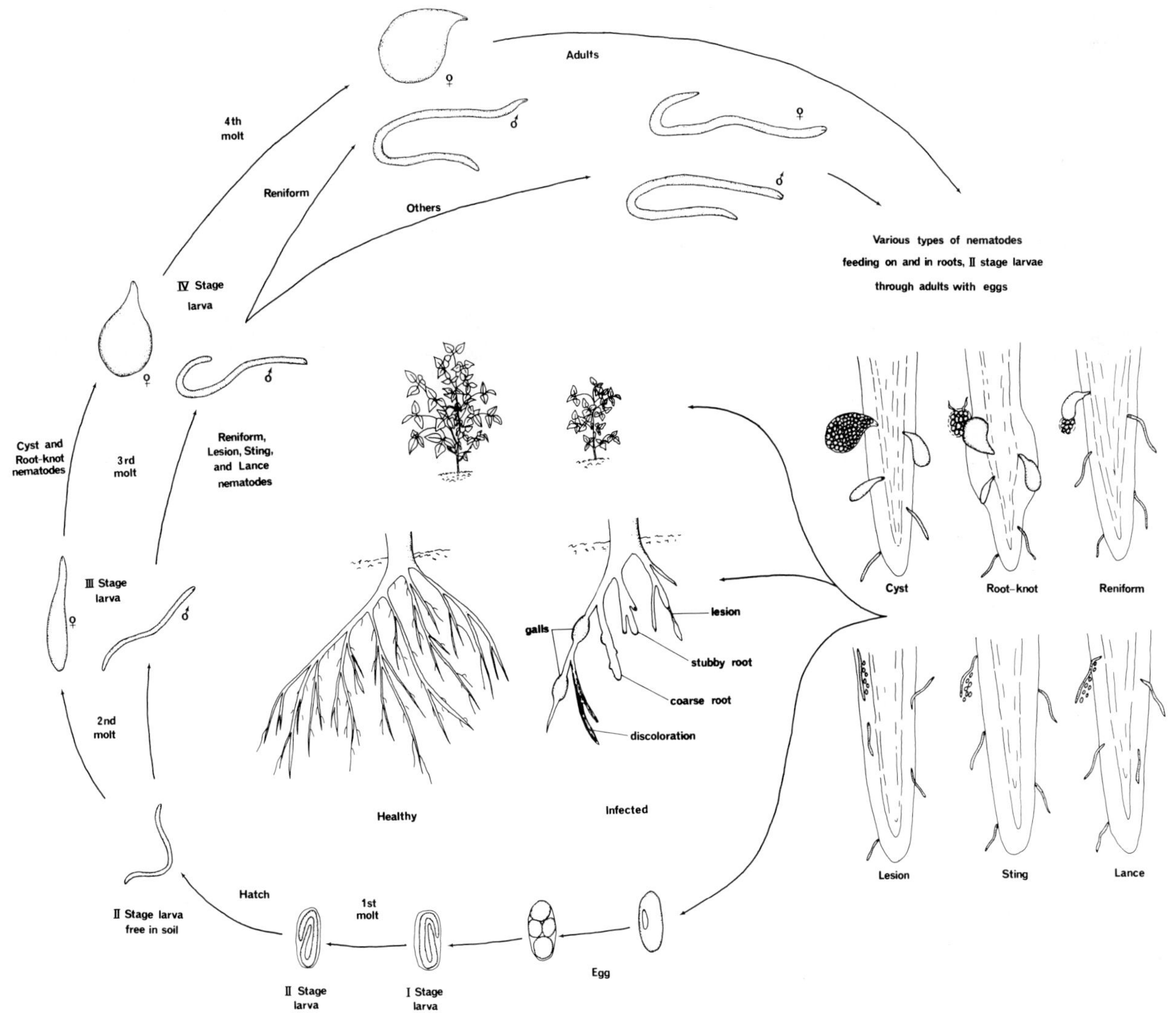

Fig. 65. Generalized life cycles of six genera of plant-parasitic nematodes that attack soybeans and symptoms they cause. (Developed by R. B. Malek; drawing by D. J. Royce)

Symptoms

Foliar symptoms of seedlings vary from slight stunting to severe chlorosis and death (Plate 27). Mature plants may be stunted or chlorotic, or both. These symptoms, however, are not diagnostic, because nitrogen and potassium deficiencies may cause similar symptoms. The root system has symptoms ranging from slight discoloration to severe necrosis. Some populations of the nematode, especially race 1, also affect nitrogen fixation. Nodulation may be slightly to completely inhibited, and the nitrogen-fixation efficiency of the remaining nodules may be reduced. Because the symptoms are misleading, diagnosis of the disease must be based on signs, namely, the white to yellow females, which erupt from the roots (Fig. 66).

Causal Organism

Cysts of *H. glycines* are lemon-shaped and 560–850 × 350–590 µm. Brown bullae (internal knobs in the anal area) are present. Young females are white when young and turn yellow with age; upon death the body wall hardens and becomes a dark brown cyst (Fig. 67). The cyst wall has a pattern of irregular, short, zigzag lines. The female produces a gelatinous matrix at the vulval cone (Fig. 67), which usually contains some eggs, and the female body is also filled with eggs (Fig. 68). Males are vermiform and 1.0–1.5 mm long. Second-stage juveniles are approximately 450 µm long. About half of the tail of the second-stage juvenile is hyaline.

H. glycines has an egg stage, four juvenile stages, and an adult stage (Fig. 69). First-stage juveniles develop within the egg and molt once to become second-stage juveniles, which emerge from the egg (in the gelatinous matrix or in the cyst). A hatching factor may stimulate hatching, but considerable spontaneous hatching occurs when eggs are not in diapause. Second-stage juveniles penetrate roots approximately 1 cm or more behind the root tip, migrate to the vascular tissue, and place their lip region adjacent to the stele. When feeding begins, the nematodes begin to enlarge and become sedentary. Three more molts occur, resulting in third- and fourth-stage juveniles and adults (males and females). Males mature faster than females and are required for reproduction. Development occurs at temperatures of 18–32°C; 24–28°C is optimal. Development does not occur above 33°C and is very slow below 16°C.

Eggs within the cyst survive for 11 years or more. Their survival may be due in part to protection by the cyst wall and in part to a genetically controlled diapause.

The soybean cyst nematode possesses a high degree of genetic variability expressed phenotypically. The ability of these variants to reproduce differentially on four soybean cultivars and lines results in a classification of 16 races (Table 10).

The relationship between soybean yield and the number of eggs is generally linear, although in some cases a tolerance limit might be expressed. This tolerance of low levels of the nematode is dependent on soil and physical factors, especially those related to moisture and nutrition.

The soybean cyst nematode can be disseminated by wind, water, soil peds in uncleaned seed (Plate 28), and machinery. Virtually anything that can move soil can disseminate this nematode.

Control

Crop rotation is effective in controlling the soybean cyst nematode, because few crops are susceptible. Growing a nonhost for two years is generally adequate to allow a susceptible soybean cultivar to be grown; however, additional benefit may be achieved by growing a nonhost for an additional year if the population density of the nematode is extremely high. A resistant cultivar may occasionally be used in place of a nonhost. In fields in which the nematode population is very low

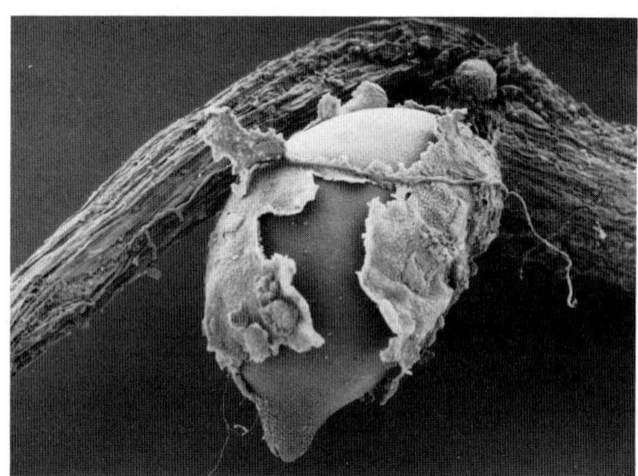

Fig. 67. Adult female soybean cyst nematode, *Heterodera glycines*, in a scanning electron micrograph (×70). (Courtesy N. M. Kokalis)

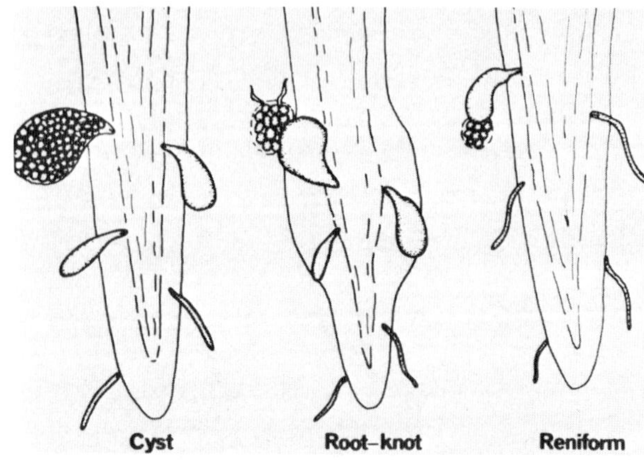

Fig. 68. Attachment and method of bearing eggs of three plant-parasitic nematodes of soybeans. (Developed by R. B. Malek; drawing by D. J. Royce)

Fig. 66. Cysts formed by female soybean cyst nematodes, *Heterodera glycines*, on soybean roots. (Courtesy R. D. Riggs and D. P. Schmitt)

or is heavily parasitized by fungi, one full year of a nonhost may be sufficient.

Cultivars with resistance to races 1, 3, and 4 are available. Most field populations appear to be mixtures of nematode genotypes. Selection forces imposed by a resistant cultivar result in changes in gene frequency of the nematode population. Continuous or frequent use of resistant cultivars results in race shifts, and "resistance-breaking" types increase.

Nematicides provide some control of this nematode. However, economics must be considered in selecting a nematicide. For example, a nematicide may increase yields enough to pay for itself, but the yield improvements may not be enough to make the crop profitable. Some nematicides may increase plant growth but not increase yields.

Long-term control of the soybean cyst nematode requires integration of control practices, including crop rotation, use of resistant cultivars, and good crop management. The particular integration of control tactics is dependent on the crop or crops, available cultivars, and the ability to manage soil water. Important considerations in integrated management are the use of nonhosts to maintain a low nematode population, good weed control (especially to control weeds that are hosts of *H. glycines*), and the use of resistant cultivars in such a way that race shifts are minimized. Nematicides may be useful if other nematode species are present or if soybean cyst nematode genotypes are mixed in such a way that some of them can be controlled with resistant cultivars but others cannot.

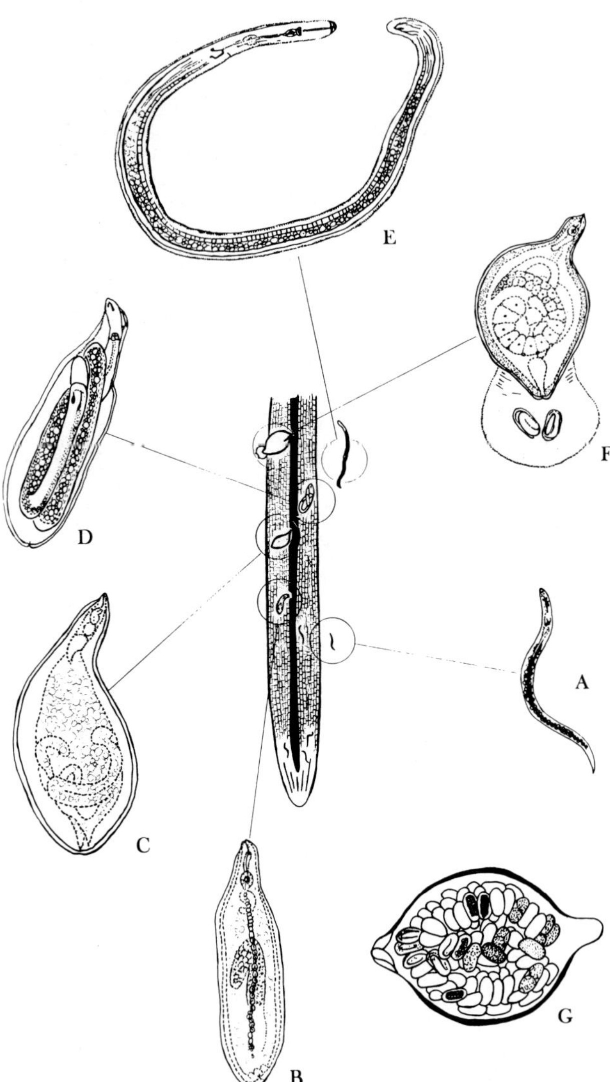

Fig. 69. Life cycle of the soybean cyst nematode, *Heterodera glycines*. **A–D,** First- through fourth-stage juveniles. **E,** Adult male. **F,** Adult female with egg sac. **G,** Cyst, containing eggs. (Courtesy R. D. Riggs and D. P. Schmitt)

Table 10. Classification of Races of *Heterodera glycines* by Means of Host Differentials

Race	Reactions of Differential Cultivars[a]			
	Pickett	Peking	PI 88788	PI 90763
1	−	−	+	−
2	+	+	+	−
3	−	−	−	−
4	+	+	+	+
5	+	−	+	−
6	+	−	−	−
7	−	+	−	−
8	−	−	−	+
9	+	+	−	−
10	+	−	−	+
11	−	+	+	−
12	−	+	−	+
13	−	−	+	+
14	+	+	−	+
15	+	−	+	+
16	−	+	+	+

[a] + = Number of females and cysts recovered is greater than or equal to 10% of the number on the cultivar Lee; − = number of females and cysts recovered is less than 10% of the number on Lee.

Selected References

Barker, K. R., Huisingh, D., and Johnston, S. A. 1972. Antagonistic interaction between *Heterodera glycines* and *Rhizobium japonicum* on soybean. Phytopathology 62:1201-1205.

Golden, A. M., Epps, J. M., Riggs, R. D., Duclos, L. A., Fox, J. A., and Bernard, R. L. 1970. Terminology and identity of infraspecific forms of the soybean cyst nematode (*Heterodera glycines*). Plant Dis. Rep. 54:544-546.

Riggs, R. D., Hamblen, M. L., and Rakes, L. 1981. Infraspecific variation in reaction to hosts in *Heterodera glycines* populations. J. Nematol. 13:171-179.

Riggs, R. D., and Schmitt, D. P. 1987. Nematodes. Pages 757-778 in: Soybeans: Improvement, Production, and Uses, 2nd ed. J. R. Wilcox, ed. American Society of Agronomy, Madison, WI.

Schmitt, D. P., Ferris, H., and Barker, K. R. 1987. Response of soybean to *Heterodera glycines* races 1 and 2 in different soil types. J. Nematol. 19:240-250.

Schmitt, D. P., and Noel, G. R. 1984. Nematode parasites of soybeans. Pages 13-59 in: Plant and Insect Nematodes. W. R. Nickle, ed. Marcel Dekker, New York.

(Prepared by R. D. Riggs and D. P. Schmitt)

Lance Nematodes

Two species of lance nematodes, *Hoplolaimus columbus* Sher and *H. galeatus* (Cobb) Thorne, attack soybeans. Infection and damage by *H. columbus* was first noted in soybeans in the early 1960s in central South Carolina, and now this species is present in cotton, maize, and soybeans in Alabama, Georgia, Louisiana, and North and South Carolina. Infection of soybean roots by *H. galeatus* occurs in the United States, but damage to soybean growth and yield is uncommon.

Symptoms

Chlorotic, stunted plants result from feeding by *H. columbus*. Damage to intolerant cultivars may occur when populations exceed 60–100 per cubic centimeter of soil, especially when moisture is limiting. In sandy soils with hardpan conditions, large numbers of secondary and tertiary roots are produced, coinciding with necrosis of the taproot.

Causal Organism

H. columbus is considered an ectoparasite, since it is exposed to the soil in all life stages. However, this species occurs in great numbers within roots (up to 500 per gram of fresh roots). The cortex is the preferred feeding site (Fig. 70), and all life stages of

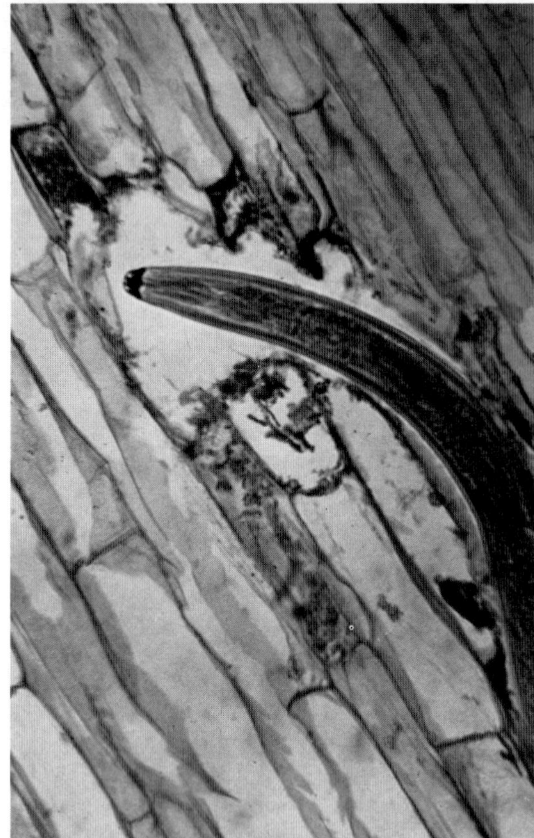

Fig. 70. Lance nematode (*Hoplolaimus columbus*) feeding in cortical tissues of a soybean root. (Courtesy S. A. Lewis)

the nematode can be found there. Older feeding sites may show evidence of vascular damage, especially in the phloem and xylem parenchyma. Most of the damage caused by this species is associated with its migration through the root tissue and the resulting lack of integrity of the cortex. Individual nematodes may be completely in or out of root tissues, whereas others stay wholly or partially embedded in roots, and still others feed ectoparasitically. Disruptions of the root epidermis where rootlets emerge or where other nematodes have penetrated are attractive to *H. columbus*. The life cycle takes about 28 days under normal conditions and occurs more rapidly at 30° C than at 25 or 20° C.

Control

Several tolerant cultivars exist in maturity groups VII and VIII, and there is some indication that cultivars tolerant to the soybean cyst nematode also have tolerance to *H. columbus*. However, tolerance to *H. columbus* has varied by location and year, and therefore precise control recommendations are difficult to make, especially since fewer acres are now treated with nematicidal chemicals. Early planting, within the recommended time for planting, is beneficial in avoiding optimal temperatures for infection and reproduction of the nematode. Some nematicidal chemicals, even at low to moderate rates, can enhance yields. Subsoiling is advised in areas with hardpan conditions, to alleviate some of the impact of nematode feeding.

Selected References

Boerma, H. R., and Hussey, R. S. 1984. Tolerance to *Heterodera glycines* in soybean. J. Nematol. 16:289-296.

Lewis, S. A., and Fassuliotis, G. 1982. Lance nematodes, *Hoplolaimus* spp., in the southern United States. Pages 127-138 in: Nematology in the Southern Region of the United States. R. D. Riggs and committee, eds. South. Coop. Ser. Bull. 276. University of Arkansas, Fayetteville.

Lewis, S. A., Smith, F. H., and Powell, W. M. 1976. Host-parasite relationships of *Hoplolaimus columbus* on cotton and soybeans. J. Nematol. 8:141-145.

Mueller, J. D., and Sanders, G. B. 1987. Control of *Hoplolaimus columbus* on late planted soybean with aldicarb. Ann. Appl. Nematol. 1:123-126.

(Prepared by S. A. Lewis)

Lesion Nematodes

Several species of lesion nematodes, *Pratylenchus* spp., attack soybeans. Lesion nematodes are found in soils throughout the world and attack a wide range of crops and weeds.

Symptoms

Lesion nematodes are endoparasitic and attack the root cortex. Their feeding generally causes dark lesions and an overall browning of the roots, which may decrease root growth by 25% (Fig. 71 and Plate 29). The epidermis and cortex of severely infected roots may slough away from the stele. The degree of damage is generally influenced by the nematode population at planting time and by temperature. If water and nutrients become limiting, plants become yellow and stunted, and yields may be reduced.

Causal Organisms

Important lesion nematodes are *P. agilis* Thorne & Malek, *P. alleni* Ferris, *P. brachyurus* (Godfrey) Filip. & Schuur.-Stek., *P. coffeae* (Zimmerman) Filip. & Schuur.-Stek., *P. hexincisus* Taylor & Jenkins, *P. penetrans* (Cobb) Filip. & Schuur.-Stek., *P. safaensis* Fortuner, *P. scribneri* Steiner, *P. vulnus* Allen & Jensen, and *P. zeae* Graham. Members of this group are among the smallest nematodes that feed on plants; the adults range from 500 to 800 μm in length. Some species, such as *P. brachyurus* and *P. zeae*, are found primarily in warm regions. Others, such as *P. hexincisus* and *P. penetrans*, are common in heavy-textured soils and are most abundant in temperate regions.

Control

Cultivars resistant or tolerant to some species of *Pratylenchus* are available in different maturity groups. Most

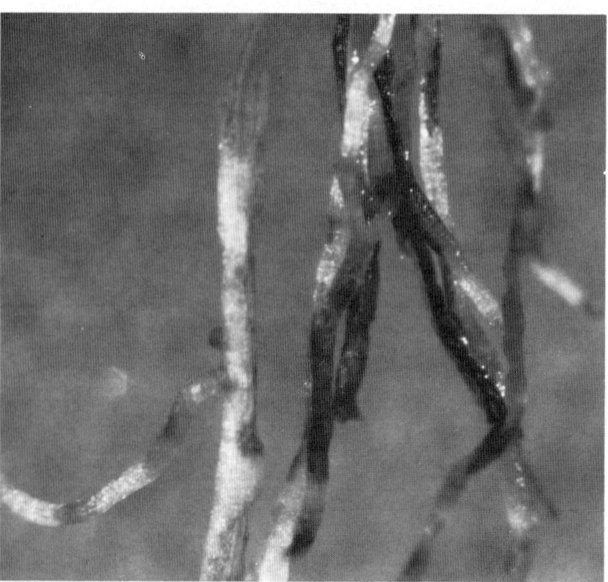

Fig. 71. Soybean roots with necrotic areas of coalesced lesions caused by the lesion nematode *Pratylenchus scribneri*. (Courtesy N. Acosta)

of these cultivars are also resistant to other nematodes, such as reniform and root-knot nematodes.

Selected References

Acosta, N. 1980. Effect of time lapse on pathogenicity and population dynamics of *Pratylenchus alleni* and *P. scribneri* in soybeans. Nematropica 10:75-80.

Acosta, N. 1982. Influence of inoculum level and temperature on pathogenicity and population development of lesion nematodes on soybean. Nematropica 12:189-197.

Acosta, N. 1982. Vertical distribution of *Pratylenchus alleni* and *P. scribneri* in soybean roots. J. Agric. Univ. P.R. 66:60-64.

Acosta, N., and Malek, R. B. 1981. Symptomology and histopathology of soybean roots infected by *Pratylenchus scribneri* and *P. alleni*. J. Nematol. 13:6-12.

Koenning, S. R., Schmitt, D. P., and Barker, K. R. 1985. Influence of selected cultural practices on winter survival of *Pratylenchus brachyurus* and subsequent effects on soybean yield. J. Nematol. 17:428-434.

Motsinger, R. E., and Minton, N. A. 1986. First report of *Pratylenchus safaensis* in the United States. Plant Dis. 70:259.

Rebois, R. V., and Golden, A. M. 1985. Pathogenicity and reproduction of *Pratylenchus agilis* in field microplots of soybeans, corn, tomato, or corn-soybean cropping systems. Plant Dis. 69:927-929.

(Prepared by N. Acosta V.)

Reniform Nematode

Infection of soybean roots by the reniform nematode *Rotylenchulus reniformis* Linford & Oliveira was reported in 1956 in the Gold Coast (West Africa) and in 1967 in South Carolina. Losses occurred only when the nematode population was high. *R. reniformis* is now distributed in the southeastern and Gulf Coast areas of the United States and most tropical regions of the world. It is thought to be incapable of overwintering in temperate latitudes but is of great potential importance to soybeans grown in the tropics.

Symptoms

Symptoms of reniform nematode infection include chlorosis, stunting, and unthrifty growth, with the production of a large number of empty pods. Galls are not produced on roots, but the root system is severely stunted and may become necrotic. The most common sign of the nematode is the soil-covered egg masses on roots.

Infection by this nematode increases the replication of cowpea chlorotic mottle virus in soybeans.

Causal Organism

R. reniformis is a semiendoparasite, partially embedding itself in roots (Fig. 72). Because it is small and because soil particles adhere to the portion of the body outside the root, the nematode is easily overlooked.

Eggs are elongate (72–100 × 33–44 µm) and hatch within 24 hr, yielding a sex ratio of about 1:1. Larvae are 330–445 µm long and have a well-developed stylet, 14–18 µm long. They molt three times in the soil with little or no root feeding and become adults. Only females parasitize plant roots. The males remain in the rhizosphere or the gelatinous matrices of the females. The females invade roots after a few days of feeding on epidermal tissues and reach permanent feeding sites in the cortex and phloem. The posterior portion of the female remains outside the root and swells to the characteristic kidney shape. The males remain vermiform. Mature females are 410–655 µm long. A female lays 70–84 eggs, outside the roots (Fig. 68). Many of the eggs are infertile.

The life cycle is completed in 17–23 days at a soil temperature of 29° C, the optimum for root invasion and nematode

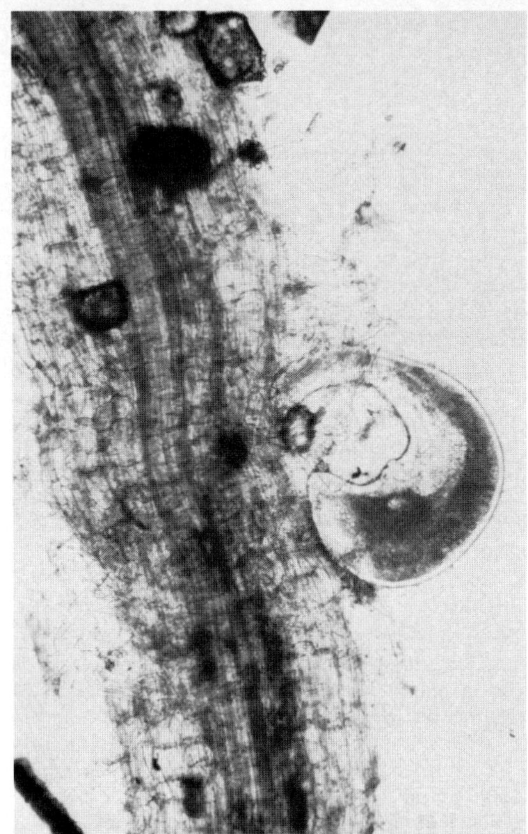

Fig. 72. Female reniform nematode (*Rotylenchulus reniformis*) with an internal egg mass, its head embedded in a soybean root. (Courtesy E. C. McGawley)

development. No root infection or egg production occurs at soil temperatures below 15° C or above 36° C. Infection is greatest when soil moisture levels are just below field capacity; extremely wet or dry conditions reduce root invasion.

Two races have been described, races A and B.

Control

Several resistant cultivars and breeding lines have been found. Resistance in soybeans is quantitative and controlled by two pairs of genes with unequal effects. A few cultivars resistant to the soybean cyst and root-knot nematodes are also resistant to the reniform nematode, although the inheritance may be separate. Production of grass crops for two or more years reduces reniform nematode populations sufficiently to allow production of a soybean crop.

Selected References

Farahat, A. A., and Kheir, A. M. 1983. Comparative histopathogenesis of certain leguminous crops infected with *Rotylenchulus reniformis*. Pak. J. Nematol. 1:57-61.

Harville, B. G., Green, A., and Birchfield, W. 1985. Genetic resistance to reniform nematodes in soybeans. Plant Dis. 69:587-589.

Lim, B. K., and Castillo, M. B. 1979. Screening soybeans for resistance to reniform nematode disease in the Philippines. J. Nematol. 11:275-282.

Rebois, R. V., Madden, P. A., and Eldridge, B. J. 1975. Some ultrastructural changes induced in resistant and susceptible soybean roots following infection by *Rotylenchulus reniformis*. J. Nematol. 7:122-139.

Vilsoni, F., and Heinlein, M. 1982. Host range and reproductive capacity of the reniform nematode *Rotylenchulus reniformis* on some crop cultivars in Fiji. Fiji Agric. J. 44:61.

(Prepared by E. C. McGawley and K. C. Hadden)

Root-Knot Nematodes

Root-knot disease of soybeans, caused by several *Meloidogyne* spp., has been reported from many soybean-growing areas of the world, though occurring only sporadically in regions primarily suitable for soybeans of maturity group V or lower-numbered groups. The incidence of the disease is greater in subtropical than in temperate regions. It is frequently considered endemic in the warm soils and long-season conditions of regions suitable for soybeans of maturity group VII or higher-numbered groups. Yield losses vary from inconsequential to complete, depending on the species and population density of the nematode, cultivar susceptibility, and intensity of drought stress to which the host is subjected. Because of the effect of drought stress, the greatest yield losses are likely to occur where soybeans are produced in sandy, light-textured soils. In addition, a high incidence of root-knot nematode infection decreases the formation of nodules by *Bradyrhizobium* spp. and increases host susceptibility to vascular pathogens.

Symptoms

The primary symptom for identifying the disease is the presence of galls on infected roots (Fig. 73). These are produced by the increased size and number of cells in infected and adjacent tissues in response to root-knot nematode feeding. The galls vary in number and size, depending on the intensity of the infection and the nematode species involved. When numerous, galls tend to coalesce, so that entire roots may be greatly swollen (Fig. 73). Vascular elements of infected tissues are disrupted, and water and nutrient flow through the plant can be inhibited, with the result that secondary symptoms of the disease are produced, including varying degrees of chlorosis and stunting and a tendency of infected plants to wilt in the heat of the day. Even in moderately infested sites plants mature 1 to 2 weeks earlier than usual. Patches of infected plants, which usually occur irregularly in the field, normally spread in the direction of cultivation and surface water flow. Infested sites frequently suffer from dense weed growth.

Causal Organisms

The species that cause root-knot disease of soybeans are *M. arenaria* (Neal) Chitwood, *M. hapla* Chitwood, *M. incognita* (Kofoid & White) Chitwood, and *M. javanica* (Treub) Chitwood.

M. hapla is the least damaging, occurring mainly in the northern reaches of the regions in which the disease occurs. Its

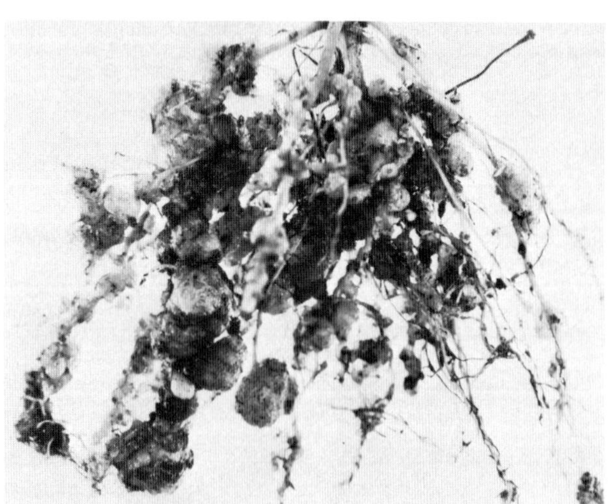

Fig. 73. Galls produced on soybean roots by a root-knot nematode, *Meloidogyne* sp. (Courtesy U.S. Department of Agriculture)

pathogenicity on soybeans is characterized by small galls and increased lateral branching of the root system.

M. javanica, which is possibly the most damaging, occurs in the most southerly areas of soybean production in the United States. It is the most common species in Brazil and other producing areas in tropical and subtropical regions. In the United States, it is found routinely on soybeans grown in areas associated with tobacco production. However, its distribution is not restricted by this association.

M. incognita and, increasingly, *M. arenaria* are the most ubiquitous agents of root-knot disease. Management of these nematodes is a necessity in much of the soybean-producing areas of the southern United States, including Alabama, Florida, Georgia, and South Carolina, and in some of the light-textured soils of Louisiana and Mississippi. Both species exist in populations of pathotypes or races that are identified by their ability to parasitize selected crop plants. This is of considerable consequence in the management of these nematodes.

Epidemiology

The disease cycle in soybeans (Fig. 74) begins with the invasion of roots, mostly at or near the root tip, by the vermiform second-stage juveniles. This can occur throughout the growing season. Migration within a root does not occur except in massive infections of the root by many juveniles. They normally settle at a feeding site close to the point of entry. Initiation of galling can be evidenced within 2 to 3 days. Following a series of three molts, the infecting nematodes become adults, a few of which are males, which are elongate (1.0–1.5 mm) and migratory. The majority are females, which are swollen and pear-shaped. Reproduction of the predominant species is by parthenogenesis. A female lays several hundred eggs, which are deposited in a gelatinous matrix through a rupture in the gall surface (Fig. 68). The developing nematodes perform one molt within the egg and hatch as second-stage infective juveniles. Several overlapping generations, each of 3–4 weeks' duration, are completed through the season.

In latitudes where soil temperatures at the time of soybean maturity are sufficient for egg hatch (above 15°C), root-knot nematodes survive the winter in soil as infective juveniles. When temperatures at soybean maturity are lower than this, they are likely to survive as eggs. The number of nematodes in the soil reaches a maximum at soybean maturity. The population declines slowly through the winter and then precipitously as soil temperatures increase in the spring. The rapid decrease in the number of infective nematodes in the soil can occur several weeks before soybean planting; it is due to a number of factors, including increased activity of the nematodes and depletion of their food reserves, along with an increase in the predatory and parasitic activities of soil arthropod and microbial communities, which are detrimental to the survival of the nematodes. By planting time, the number of nematodes in the soil is at its lowest, usually less than 10% of that at soybean maturity.

Control

Damaging levels of root-knot nematodes in soybeans usually occur following several years of continuous planting of the crop or other hosts. Consequently, both preventive and corrective management is needed to avoid this problem. In addition, planting and cultivating equipment must be washed free of contaminated soil, and fields should be maintained free of weeds, many of which are hosts of root-knot nematodes. Nematicides are rarely profitable for controlling nematodes in soybeans. Consequently, growers must rely on a good selection of crop systems for effective management of root-knot disease of soybeans. Though *Meloidogyne* spp. have wide host ranges, advantage can be taken of the differing abilities of the various hosts to affect population levels of the nematodes. Graminaceous crops are less susceptible hosts of the various root-knot nematode species than soybeans and, when grown in rotation with soybeans, have a controlling effect on residual populations of juveniles. When soil infestation levels of juveniles reach damaging densities, summer planting with a graminaceous crop is beneficial, irrespective of the *Meloidogyne* sp. present.

Several soybean cultivars in maturity groups V through VIII are tolerant to some root-knot nematode species and should be grown, even in soils not suspected of harboring the nematodes. Resistance to root-knot disease is horizontal in soybeans, and some yield loss occurs in these cultivars when populations of root-knot nematodes in the soil are dense.

Because of parthenogenetic reproduction, natural selection of resistance-breaking populations of root-knot nematodes on soybeans has not been experienced. However, since infested sites may harbor more than one species of root-knot nematode, a species shift due to selective crop and cultivar planting may occur.

Where root-knot disease is recognized, accurate identification of the causal species and their pathotypes allows greater flexibility in the choice of cropping systems available for management. Knowledge of the causal species is mandatory for the selection of the correct soybean cultivar.

Selected References

Kinloch, R. A. 1980. The control of nematodes injurious to soybean. Nematropica 10:141-153.

Riggs, R. D., and committee, eds. 1982. Nematology in the Southern Region of the United States. South. Coop. Ser. Bull. 276. University of Arkansas, Fayetteville. 206 pp.

Rodríguez-Kábana, R., Robertson, D. G., King, P. S., and Weaver, C. F. 1987. Evaluation of nematicides for control of root-knot and cyst nematodes on a tolerant soybean cultivar. Nematropica 17:61-70.

Rodríguez-Kábana, R., and Weaver, D. B. 1984. Soybean cultivars and development of populations of *Meloidogyne incognita* in soil. Nematropica 14:46-56.

Sasser, J. N., and Carter, C. C., eds. 1985. An Advanced Treatise on *Meloidogyne*, Vols. I and II. North Carolina State University, Raleigh.

Schmitt, D. P. 1985. Plant-parasitic nematodes associated with soybeans. Pages 541-546 in: Proc. World Soybean Res. Conf. III. R. Shibles, ed. Westview Press, Boulder, CO.

Trivedi, P. C., and Barker, K. R. 1986. Management of nematodes by cultural practices. Nematropica 16:213-236.

(Prepared by R. A. Kinloch and R. Rodríguez-Kábana)

Sting Nematodes

The sting nematodes *Belonolaimus gracilis* Steiner and *B. longicaudatus* Rau have a wide host range, including soybeans. Potential yield losses range up to 50%. The nematodes are

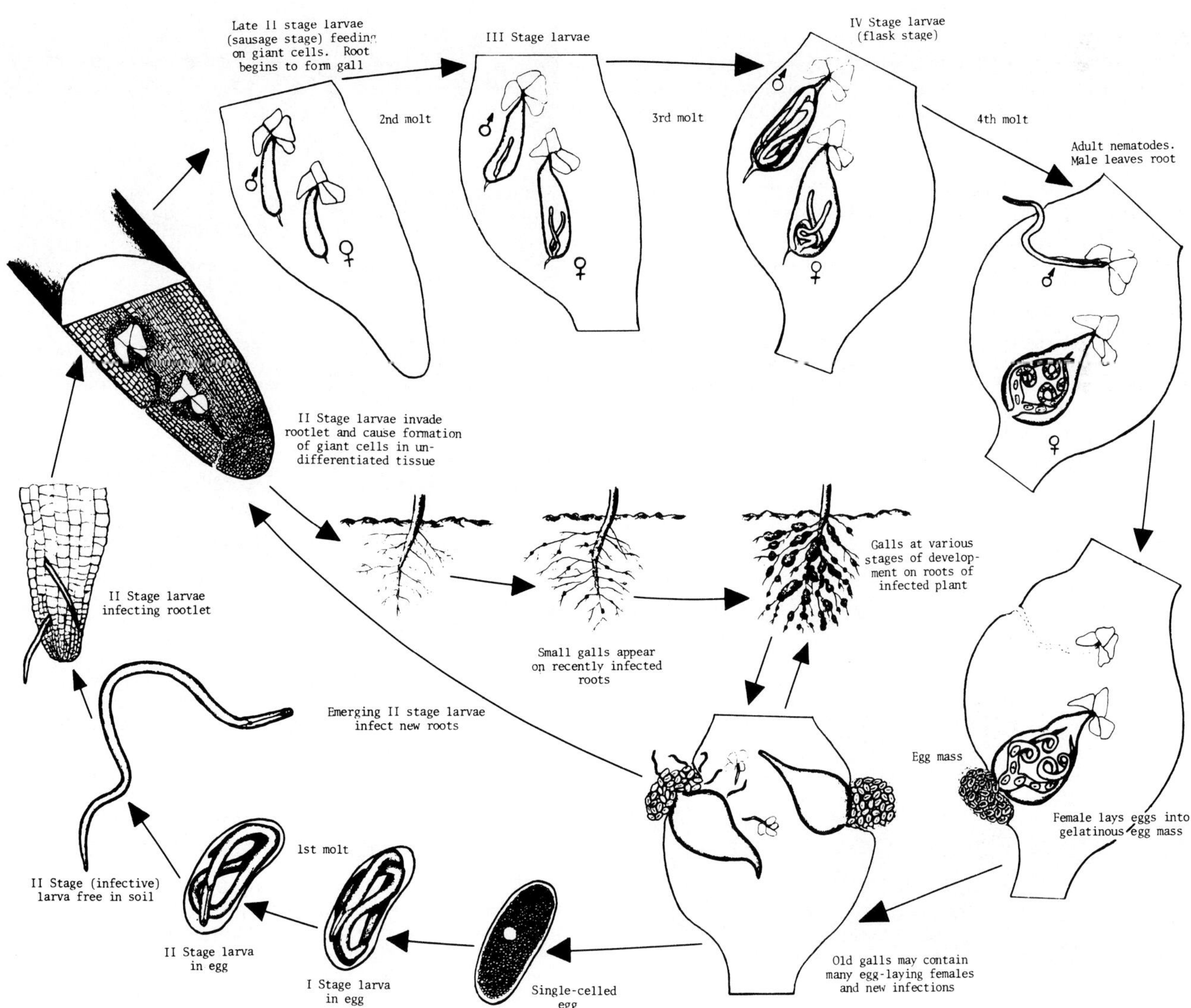

Fig. 74. Life cycle of a root-knot nematode, *Meloidogyne* sp., the cause of root knot of soybeans. (Adapted from G. N. Agrios, 1988, Plant Pathology, 3rd ed., Academic Press, New York)

found primarily in the southern Atlantic seaboard; infestations also occur in Alabama, Arkansas, Kansas, New Jersey, Oklahoma, and Texas. They are most common in sandy soils.

The initial symptoms are minute, dark, sunken lesions along the root axis or at root tips. These lesions may enlarge to girdle the roots or extend along the longitudinal axis of affected roots. Terminal growth often ceases, meristematic tissues are destroyed, and roots usually break off at girdling sites, so that they have a stubby-root appearance. Root proliferation is stimulated above points of attack, resulting in the typical many-branched root system of plants parasitized by sting nematodes. Top growth is chlorotic and severely stunted, as in many other diseases.

Other Nematodes

In addition to the cyst, lesion, reniform, and root-knot nematodes, over 20 other genera of plant-parasitic nematodes have been reported in soybean fields. Whether all of the reported species can reproduce on soybeans, however, is uncertain. In some cases, a species may cause serious yield losses on other crops, but little is known about its relationship to soybeans. In the field, symptoms of damage caused by these nematodes cannot be differentiated from symptoms caused by other pathogens.

The following have been found associated with soybeans: *Criconemella* spp.; *C. simile* (Cobb) Chitwood; *Helicotylenchus delhiensis* Khan & Nanjappa; *H. dihystera* (Cobb) Sher; *H. paragirus* Saha, Chawla, & Khan; *H. pseudorobustus* (Steiner) Golden; *Paratylenchus projectus* Jenkins; *Pratylenchus crenatus* Loof; *Rotylenchus* spp.; *Scutellonema cavenessi* Sher; *Trichodorus* spp.; *Tylenchorhynchus agri* Ferris; *T. capitatus* Allen; *T. claytoni* Steiner; *T. martini* Fielding; *T. nudus* Allen; *T. vulgaris* Upadhyay, Swarup, & Sethi; *Xiphinema americanum* Cobb; and *X. diversicaudatum* (Micoletsky) Thorne. Their effects on plant growth and yield are unknown.

Seed Pathology

More than 40 fungi, bacteria, and viruses, of which 15 are of major economic importance, are associated with soybean seeds. Microorganisms and viruses that invade and colonize seeds before harvest can reduce the yield and quality of the seed crop (Fig. 75). The discoloration of seeds caused by microorganisms and viruses can limit the acceptance of seed lots, reduce the grade, and lower the prices that growers receive (Fig. 76).

Seed infection can act as a means of survival for pathogens; infected or infested seeds provide inoculum that may infect the new crop when the seeds are planted. Pathogens and nonpathogens may be disseminated over long distances and introduced into new areas via infected or infested seeds.

Some pathogens produce distinct symptoms on infected seeds (Fig. 77); however, lack of symptoms does not mean that seeds are free of pathogens. Moreover, more than one pathogen can infect the same seeds, resulting in a variety of symptoms. Root-rot fungi, such as *Fusarium* spp., *Macrophomina phaseolina* (Tassi) Goid., *Rhizoctonia solani* Kühn, *Sclerotinia sclerotiorum* (Lib.) d By., and *Sclerotium rolfsii* Sacc., can infect seeds, particularly when pods are in contact with infested soil. Infection rates of *Fusarium* spp. and *M. phaseolina* have reached 50%, but *R. solani* and *S. rolfsii* are encountered in less than 10% of any seed lot.

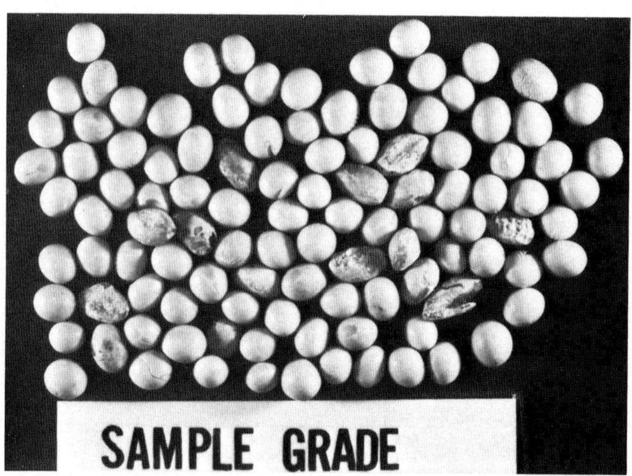

Fig. 76. Soybean seeds representing sample grade by grain standards of the U.S. Department of Agriculture. This grade is the lowest of five and receives the lowest price at the market. (Courtesy P. R. Hepperly)

Fig. 75. Fungi and bacteria growing from surface-sterilized soybean seeds plated on potato-dextrose agar, indicating the variety of internally seedborne microorganisms. (Courtesy M. A. Ellis)

Fig. 77. Soybean seed lots with symptoms produced by seed-borne pathogens (clockwise from upper right): seeds with a light brown hilum ring from soybean mosaic virus infection; seeds infected with *Cercospora sojina*, the frogeye leaf spot fungus; seeds infected with *C. kikuchii*, the purple seed stain fungus; seeds infected with *Phomopsis longicolla*, the Phomopsis seed decay fungus; seeds with a black hilum ring from soybean mosaic virus infection; seeds without symptoms. (Courtesy J. T. Yorinori)

Table 11. Appearance of Various Fungi in Soybean Seed Coat Tissues in Thin Section Prepared for Bright-Field Microscopy[a]

Fungus	Hyphal Width (μm)	Unstained Mature Hyphae	Hyphal Aggregate	Mycelial Mat	Sclerotia	Oil Globule
Alternaria alternata	1.8–5.4	Brown	No	Yes	No	No
Cercospora kikuchii	1.3–3.0	Light brown	Yes	No	No	No
C. sojina	1.3–2.7	Brown	Yes	Yes	No	No
Colletotrichum truncatum	3–11	Brown	No	Yes	No	Yes[b]
Fusarium spp.	1.4–3.6	Hyaline	No	No	No	No
Macrophomina phaseolina	1.8–4.5	Brown	No	No	Yes	No
Phomopsis spp.	3.8–8.7	Hyaline	Yes	Yes	No	Yes

[a] Data from Singh and Sinclair, 1985, and Kunwar et al, 1985.
[b] Prominent oil globules.

The appearance of various fungi in soybean seed coat tissues prepared for bright-field microscopy is summarized in Table 11.

General References

Agarwal, V. K., and Sinclair, J. B. 1987. Principles of Seed Pathology. CRC Press, Boca Raton, FL. 2 vols., 244 pp.

Bowman, J. E., Hartman, G. L., McClary, R. D., Sinclair, J. B., Hummel, J. W., and Wax, L. M. 1986. Effects of weed control and row spacing in conventional tillage, reduced tillage, and nontillage on soybean seed quality. Plant Dis. 70:673-676.

Hill, H. J., and West, S. H. 1982. Fungal penetration of soybean seed through pores. Crop Sci. 22:602-605.

Kunwar, I. K., Singh, T., and Sinclair, J. B. 1985. Histopathology of mixed infections by *Colletotrichum truncatum* and *Phomopsis* spp. or *Cercospora sojina* in soybean seeds. Phytopathology 75:489-492.

McGee, D. C., Brandt, C. L., and Burris, J. S. 1980. Seed mycoflora of soybeans relative to fungal interactions, seedling emergence, and carry over of pathogens to subsequent crops. Phytopathology 70:615-617.

Ndimande, B. N., Wein, H. C., and Kueneman, E. A. 1981. Soybean seed deterioration in the tropics. I. The role of physiological factors and fungal pathogens. Field Crops Res. 4:113-121.

Richardson, M. J. 1979. An Annotated List of Seed-Borne Diseases, 3rd ed. Phytopathol. Pap. 23. Commonwealth Mycological Institute, Kew, Surrey, England. 320 pp. Suppl. 1, 1981, 78 pp. Suppl. 2, 1983, 108 pp.

Sinclair, J. B. 1986. Multiple fungal infections of soybean seeds in preharvest and postharvest deterioration. Pages 65-76 in: Physiological-Pathological Interactions Affecting Seed Deterioration. S. H. West, ed. Crop Science Society of America, Madison, WI.

Sinclair, J. B., and Jackobs, J. A., eds. 1981. Soybean Seed Quality and Stand Establishment. INTSOY Ser. 22. College of Agriculture, University of Illinois at Urbana-Champaign. 240 pp.

Singh, T., and Sinclair, J. B. 1985. Histopathology of *Cercospora sojina* in soybean seeds. Phytopathology 75:185-189.

(Prepared by M. M. Kulik and J. B. Sinclair)

Detection of Seedborne Pathogens

Soybean seedborne fungi are usually detected by placing surface-sterilized seeds on a nutrient agar such as potato-dextrose agar (Fig. 75) or on moist blotters (Fig. 78). After a suitable period of incubation, usually under longwave near-ultraviolet light to stimulate sporulation, the seeds are examined individually under a dissecting microscope for the presence of characteristic fungal fruiting bodies. In addition, examination of conidia under a bright-field compound microscope may be necessary.

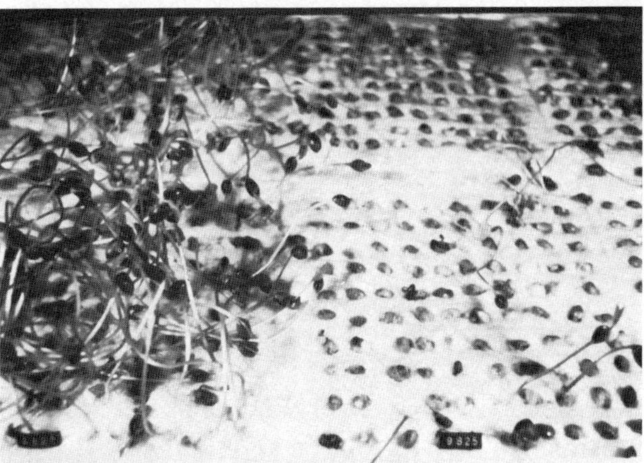

Fig. 78. Soybean seed lots germinated on moist cellulose pads (blotters), with germination and seedling vigor reduced by seedborne microorganisms. (Courtesy A. L. Lang)

Seedborne bacteria may also be detected by placing seeds on agar or on blotters. Growing-out tests may be useful for detecting seedborne viruses. New techniques for the detection of seedborne bacteria and viruses have been developed in recent years. These include enzyme-linked immunosorbent assay (ELISA), radioimmunosorbent assay (RISA), solid-phase radioimmunoassay (SPRIA), and serologically specific electron microscopy (SSEM), all of which are serological methods. The last three of these require complicated laboratory equipment, but all of them are much faster to run than the agar or blotter health tests.

Selected References

Kulik, M. M. 1984. New techniques for the detection of seed-borne pathogenic viruses, bacteria, and fungi. Seed Sci. Technol. 12:831-840.

Limonard, T. 1968. Ecological Aspects of Seed Health Testing. International Seed Testing Association, Wageningen, Netherlands. 167 pp.

Naumova, N. A. 1972. Testing of Seeds for Fungus and Bacterial Infections, 3rd ed. U.S. Department of Agriculture and National Science Foundation, Washington, D.C. 145 pp.

Phatak, H. C. 1974. Seed-borne plant viruses—Identification and diagnosis in seed health testing. Seed Sci. Technol. 2:3-155.

(Prepared by M. M. Kulik and J. B. Sinclair)

Seedborne Bacteria and Bacterial Diseases of Seeds

Bacillus Seed Decay

Bacillus seed decay is ubiquitous. It causes losses under hot (25–35°C), moist conditions in storage, in the field, and in experimental situations. Unconfirmed field losses of nearly 100% were reported in Nigeria when soybean seeds were planted in wet soil above 30°C. The disease frequently appears on seeds being tested under warm-germination conditions (e.g., accelerated aging tests). The epidemiology of the disease has not been studied.

Symptoms

In the field, in storage, or under warm-germination conditions, the bacterium causes a soft decay of seeds (Fig. 79).

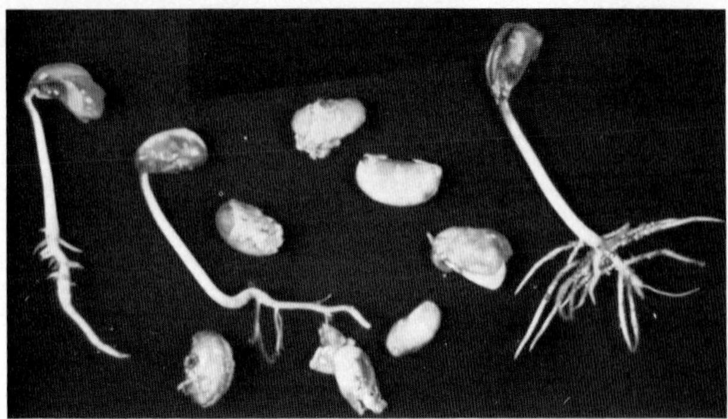

Fig. 79. Soybean seeds and seedlings in various stages of decay caused by *Bacillus subtilis*. (Courtesy J. F. Nicholson)

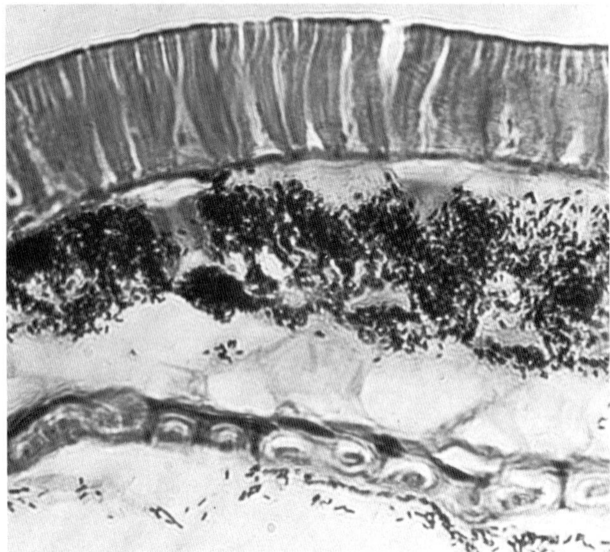

Fig. 80. Stained cells of *Bacillus subtilis* in the hourglass cell layer of the seed coat of an asymptomatic soybean. (Courtesy of S. R. Foor)

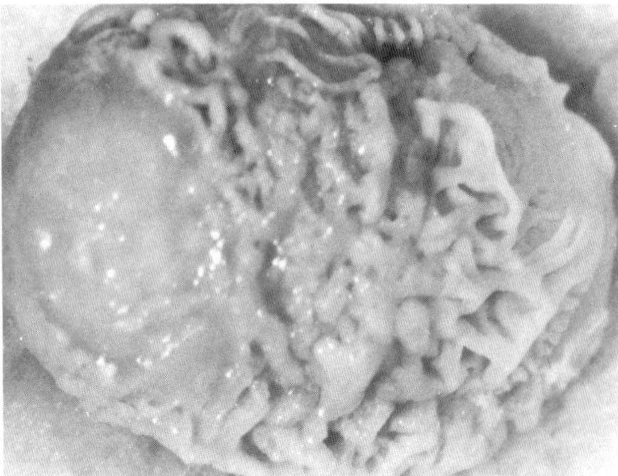

Fig. 81. Soybean seed covered with growth of *Bacillus subtilis*. (Courtesy F. D. Tenne)

The decay increases with increases in moisture and temperature. In the laboratory at temperatures of 30° C or higher and relative humidity approaching 100%, decay occurs within 5 days. Seeds are colonized at first in the seed coat, without conspicuous symptoms (Fig. 80). The bacterium produces rough or smooth, slimy, glistening colonies on potato-dextrose agar and soybean seeds (Fig. 81).

Causal Organism

Bacillus seed decay is caused by *Bacillus subtilis* (Ehrenberg) Cohn, a soil resident and an epiphyte on soybean leaves and other plant tissues. The bacterium is a motile, gram-positive, aerobic rod ($0.7-0.8 \times 2.0-3.0 \, \mu$m) with peritrichous flagellae. It forms endospores. Colonies on nutrient agar are white to cream-colored, wrinkled or smooth, and folded. The minimum and maximum temperatures for growth are 5 and 55° C. Culture filtrates of the bacterium were found antagonistic to seven soybean pathogens in culture.

Control

1. Use an antibiotic seed treatment.
2. Use a mulch at planting time for soil of high moisture content and above 30° C.

Selected References

Cubeta, M. A., Hartman, G. L., and Sinclair, J. B. 1985. Interaction between *Bacillus subtilis* and fungi associated with soybean seeds. Plant Dis. 69:506-509.

Ellis, M. A., Tenne, F. D., and Sinclair, J. B. 1977. Effect of antibiotics and high temperature storage on decay of soybean seeds by *Bacillus subtilis*. Seed Sci. Technol. 5:753-761.

Schiller, C. T., Ellis, M. A., Tenne, F. D., and Sinclair, J. B. 1977. Effect of *Bacillus subtilis* on soybean seed decay, germination, and stand inhibition. Plant Dis. Rep. 61:213-217.

Tenne, F. D., Foor, S. R., and Sinclair, J. B. 1977. Association of *Bacillus subtilis* with soybean seeds. Seed Sci. Technol. 5:763-769.

(Prepared by M. M. Kulik and J. B. Sinclair)

Other Seedborne Bacteria

Stored seeds infected with *Pseudomonas syringae* pv. *glycinea* (Coerper) Young, Dye, & Wilkie (see Bacterial Blight) may shrivel, develop sunken or raised lesions, and become slightly discolored (Plate 2), or they may be symptomless. In high-moisture conditions, a slimy growth may appear on the surface of infected seeds.

Other bacteria that are seedborne in soybeans include *Curtobacterium flaccumfaciens* pv. *flaccumfaciens* (Hedges) Collins & Jones (see Bacterial Tan Spot); *P. solanacearum* (Smith) Smith (see Bacterial Wilts); *P. syringae* pv. *tabaci* (Wolf & Foster) Young, Dye, & Wilkie (see Wildfire); and *Xanthomonas campestris* pv. *glycines* (Nakano) Dye (see Bacterial Pustule).

Selected Reference

Fett, W. F. 1979. Survival of *Pseudomonas glycinea* and *Xanthomonas phaseoli* var. *sojensis* in leaf debris and soybean seed in Brazil. Plant Dis. Rep. 63:79-83.

(Prepared by M. M. Kulik and J. B. Sinclair)

Seedborne Fungi and Fungal Diseases of Seeds

Alternaria Pod and Seed Decay

Alternaria spp. have been associated with all parts of soybean plants, everywhere the crop is grown (see Alternaria Leaf Spot), but have been described in soybean pods and seeds only in mixed infections associated with senescence, frost injury, insect damage, or wounding (Fig. 82).

Causal Organisms

A. tenuissima (Kunze ex Pers.) Wiltshire and *A. alternata* (Fr.) Keissler (syn. *A. tenuis* C. G. Nees) cause pod and seed decay. *Alternaria* spp. are easily identified by their conidia. Those of *A. tenuissima* are smooth and pale brown to golden, with short or long beaks and sometimes a terminal beak swelling. They are formed in short, unbranching chains of eight or fewer. Conidia of *A. alternata* are pale to dark brown, smooth, echinulate, and short-beaked, with no terminal beak swelling. They are sometimes formed in branching chains of eight or more.

The appearance of *A. alternata* in soybean seed coats as examined in histopathological procedures is summarized in Table 11.

Fig. 82. Soybean seed lot with smaller than normal, shriveled, discolored seeds injured by the bean leaf beetle and infected with *Alternaria tenuissima*. (Courtesy B. J. Shortt)

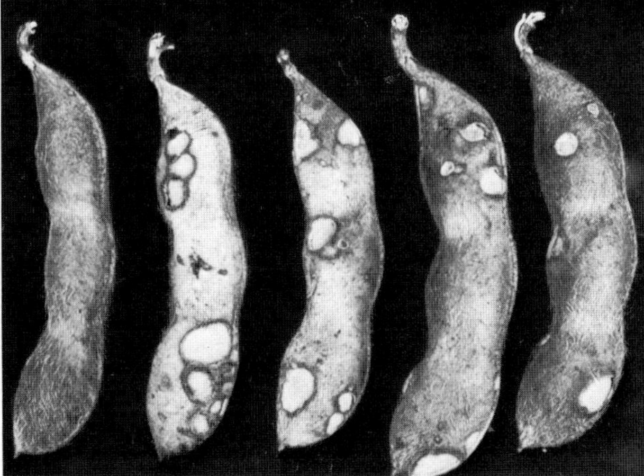

Fig. 83. Bean leaf beetle (*Cerotoma trifurcata*) feeding on soybean tissue (above); soybean pods injured by the beetle (below), with an uninjured pod (extreme left). (Courtesy B. J. Shortt)

Epidemiology

In Illinois in 1980 and 1984, the incidence of seedborne *A. tenuissima* was correlated with pod injury caused by abnormally high populations of the bean leaf beetle, *Cerotoma trifurcata* (Forster) (Fig. 83). As a result, seed viability was reduced. High levels of *Alternaria*-infected seed have frequently been observed after harvest when soybeans were treated with benomyl during the growing season. *A. alternata* was also observed on seeds infected with *A. tenuissima*. In a greenhouse study, *A. tenuissima* decayed seeds within injured pods but not within uninjured pods.

Symptomatic seeds naturally colonized by *A. alternata* had lower volume, lower density, and more splits than seeds without symptoms.

Control

1. Control insects.
2. Avoid injury to soybean pods.

Selected References

Anderson, T. R. 1987. Pod necrosis: A new disease affecting soybean on droughty soils in southwestern Ontario. (Abstr.) Annu. Meet. Can. Phytopathol. Soc., Ottawa, p. 142.

Kunwar, I. K., Manandhar, J. B., and Sinclair, J. B. 1986. Histopathology of soybean seeds infected with *Alternaria alternata*. Phytopathology 76:543-546.

Shortt, B. J., Sinclair, J. B., Helm, C. G., Jeffords, M. R., and Kogan, M. 1982. Soybean seed quality losses associated with bean leaf beetles and *Alternaria tenuissima*. Phytopathology 72:615-618.

(Prepared by M. M. Kulik and J. B. Sinclair)

Purple Seed Stain

Purple seed stain—also known as purple blotch, purple speck, purple spot, or lavender spot—was first reported in Korea in 1921 and in the United States in 1924. It is now found worldwide. The disease does not reduce yields directly, but sometimes a high percentage of seeds is stained at harvest. The greater the area of discoloration exceeding 50% of the surface, the greater the possibility of delayed germination. Seeds discolored over nearly 100% of the surface tend to have a lower oil content and a higher protein content than seeds that are not discolored.

Symptoms

Soybean seeds, pods, stems, and leaves can be infected (see Cercospora Blight and Leaf Spot). The disease is most conspicuous and easily distinguished on seeds. Seed discoloration varies from pink or pale purple to dark purple (Figs. 77 and 84), and the discolored areas range from specks to large, irregular blotches, which may cover the entire surface of the seed coat (Plates 19, 30, and 31). Cotyledons are generally not discolored. Seeds that carry the pathogen may not show symptoms.

Heavily infected seeds can produce diseased seedlings and reduced stands. Infected cotyledons often shrivel, turn dark purple, and fall prematurely. Cotyledon infection may spread to the stem, producing necrotic areas that girdle the stem and kill the young plant. Less severely affected seedlings survive and are stunted. If the weather is humid and warm during seedling emergence, a velvety, grayish white growth of conidiophores and conidia appears on infected portions of cotyledons and stems.

Causal Organism

Cercospora kikuchii (T. Matsu. & Tomoyasu) Gardner (syn. *Cercosporina kikuchii* T. Matsu. & Tomoyasu), the causal fungus, sporulates abundantly at 23–27° C within 3–5 days on infected plant tissues, including excised seed coats, as well as on laboratory media such as carrot leaf decoction agar, dead soybean plant tissue agar, potato-dextrose agar, and V-8 juice agar when the agar plates are inoculated with mycelial fragments and incubated in alternating 12-hr periods of light and darkness. Conidiophores are in fascicles, with several to more than 20 stalks on a stroma. The conidiophores are yellowish brown to dark brown at the base, gradually lighter in color up the stalk, and hyaline at the tip (Figs. 22 and 85). Conidial scars are not found on newly formed conidiophores but are evident on old conidiophores (Fig. 86). The scars are 10–150 μm apart.

A holoblastic conidium is produced at the tip of the conidiophore (Fig. 86). As the conidium matures, a cross-wall forms at the attachment. Mature conidia detach easily, and the conidiophores proliferate sympodially to form additional conidiogenous loci at the new apexes.

The conidia are hyaline, acicular truncate, straight or curved, and multiseptate, with a thickened hilum (Figs. 22 and 86). They may develop in an alternating manner, but this is not a persistent feature. They germinate in distilled water, producing one or many germ tubes within 2 or 3 hr (Fig. 86). Conidiophores and conidia in Japan are 4–6 × 85–200 μm and 4–5 × 70–164 μm, respectively, with zero to 22 septations; in North Carolina, they are 3.2–6 × 36–286 μm and 1.3–61.1 × 38.8–445 μm, respectively, with two to 49 septations.

Fig. 84. Soybean seed lot with symptoms of purple seed stain (left), caused by *Cercospora kikuchii*, and an uninfected seed lot (right). (Courtesy U.S. Department of Agriculture)

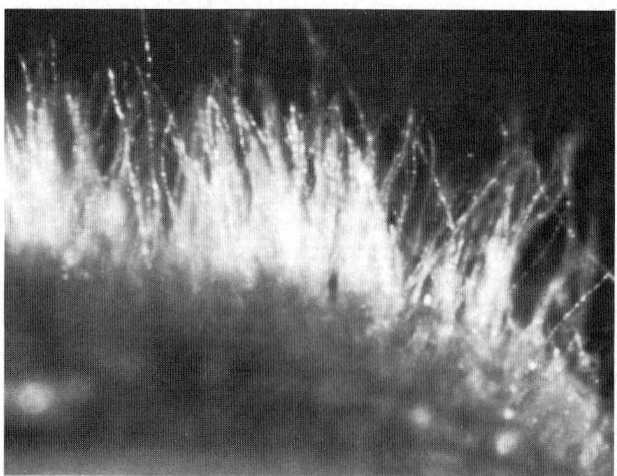

Fig. 85. Sporulating *Cercospora kikuchii*, with hyaline conidia and darker conidiophores, in a bright-field photomicrograph. (Courtesy J. T. Yorinori)

Environmental factors such as relative humidity and the composition and pH of the substrate influence the development and morphology of *C. kikuchii*. The optimum temperature for sporulation on infected seeds plated on moist filter paper is between 23 and 27°C (the minimum is 18°C). The optimum, maximum, and minimum temperatures for growth on potato-dextrose agar are 27, 35, and 12°C, respectively. The optimum pH for growth in culture is 5.9.

C. kikuchii grows well on carrot, oatmeal, and potato-dextrose agars. Growth is uniformly dense, with deep folds radiating from the center. Colonies are white at the edges and light grayish olive toward the center. The medium beneath the colony varies in color but is often dark purple with pink margins. The pigment produced on agar can be confused with that produced by *Chaetomium cupreum* Ames.

Young hyphae in culture are hyaline, septate, 2–4 μm thick, granular, and sometimes noduled. Older hyphae are pale brown, 3–5 μm thick, and closely septate. Orange-brown, thick-walled cells that resemble chlamydospores, 6–15 μm in diameter, form in old cultures.

The appearance of the fungus in soybean seed coats prepared for histopathological procedures is summarized in Table 11.

Other *Cercospora* spp., when introduced into soybean pods, cause a purple stain indistinguishable from that produced by *C. kikuchii*. The ability of these species to invade injured pods and their prevalence under natural conditions have not been determined.

Disease Cycle and Epidemiology

C. kikuchii overseasons in diseased leaves, stems, and seeds and infects soybean plants at flowering, earlier than many seedborne pathogens. Unlike other seedborne pathogens, it does not increase in incidence with delayed harvest. Several crops and weeds may serve as alternate hosts.

When infected seeds are planted, the fungus grows from the seed coat into the cotyledons, gradually extending toward the radicle and rootlets. The seed coat of lightly affected seeds generally slips off before emergence, and seedlings escape infection. The seed coat of severely affected seeds, however, may adhere to the emerging cotyledons and grow into them. Infected cotyledons remain attached or eventually fall. In either case, the fungus grows into the stem of only a small percentage of seedlings. It can establish symptomless colonization, which can persist far into the season.

In warm, humid weather, the fungus grows on cotyledons, stems, and leaves and produces conidiophores and conidia. The latter are borne by wind and splashing water to other leaves and stems, where they initiate secondary infections, which may be symptomless. Alternatively, they may produce conidia that infect other plants under warm, wet conditions.

The fungus grows through the mesocarp parenchyma of pods and then through the adaxial vein. From there, it spreads through the hilum and into the seed coat, where it produces the characteristic purple stain.

Control

1. Plant high-yielding, moderately resistant cultivars.
2. Sow high-quality seeds relatively free of the pathogen on the late side of the planting season.
3. Treat seeds with a fungicide. This helps to prevent seedling infection.
4. Apply a benzimidazole fungicide at 60% of full pod set.
5. Rotate soybeans with a nonleguminous crop.
6. Plow under crop residues.

Selected References

Chen, M. D., Lyda, S. D., and Halliwell, R. S. 1979. Infection of soybeans with conidia of *Cercospora kikuchii*. Mycologia 71:1158-1165.

Hepperly, P. R., and Sinclair, J. B. 1981. Relationships among *Cercospora kikuchii*, other seed mycoflora, and germination of

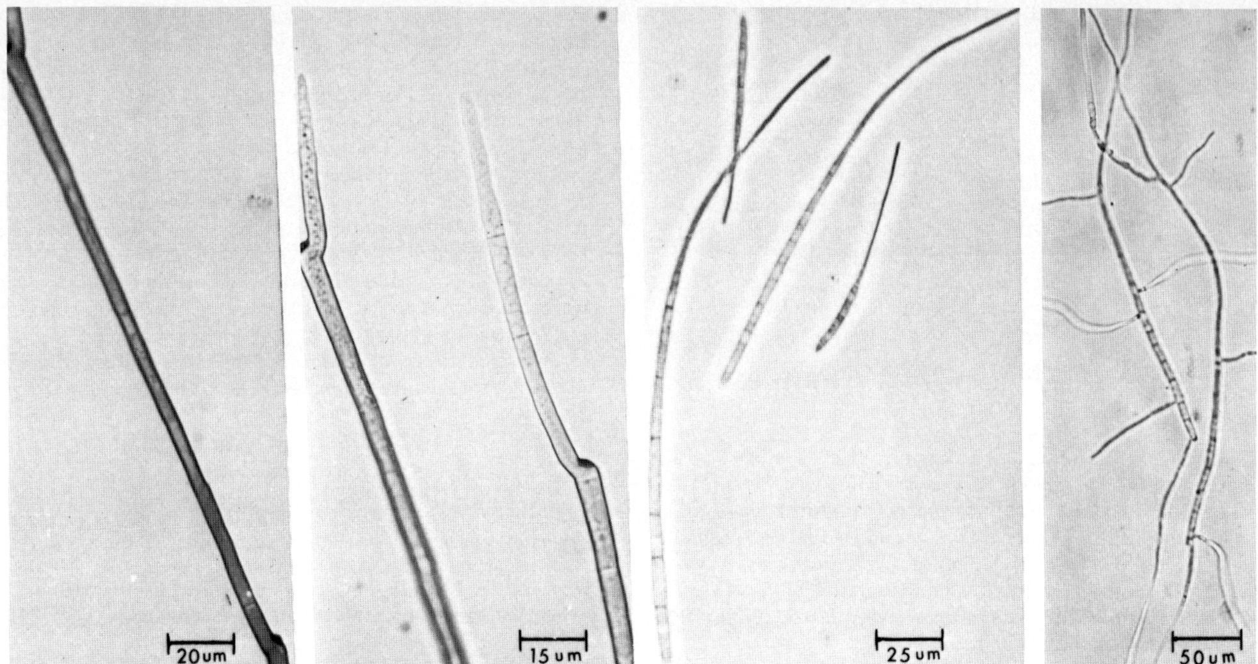

Fig. 86. *Cercospora kikuchii*, the cause of purple seed stain of soybeans (left to right): conidiophore with alternate arrangement of conidial scars; sympodial development of conidiophores; mature conidia; germinating conidia. (Courtesy C. C. Yeh)

soybeans in Puerto Rico and Illinois. Plant Dis. 65:130-132.

Roy, K. W. 1982. *Cercospora kikuchii* and other pigmented *Cercospora* species: Cultural and reproductive characteristics and pathogenicity to soybean. Can. J. Plant Pathol. 4:226-232.

Singh, D. P., and Agarwal, V. K. 1986. Purple stain of soybean and seed viability. Seed Res. 14:126.

Singh, T., and Sinclair, J. B. 1986. Further studies on the colonization of soybean seeds by *Cercospora kikuchii* and *Phomopsis* sp. Seed Sci. Technol. 14:71-77.

Vathakos, M. G., and Walters, H. J. 1979. Production of conidia by *Cercospora kikuchii* in culture. Phytopathology 69:832-833.

Yeh, C. C., and Sinclair, J. B. 1980. Sporulation and variation in size of conidia and conidiophores among five isolates of *Cercospora kikuchii*. Plant Dis. 64:373-374.

(Prepared by M. M. Kulik and J. B. Sinclair)

Cercospora sojina

Seeds infected with *Cercospora sojina* Hara develop conspicuous light to dark gray or brown areas, which vary from specks to large blotches covering the entire seed coat (Fig. 77 and Plates 19 and 30). The greater the area of discoloration, the greater the delay of germination. Some lesions show alternating bands of light and dark brown. Brown and gray lesions occasionally diffuse into each other. Usually, the seed coat cracks or flakes (see Frogeye Leaf Spot). The appearance of the fungus in soybean seed coats prepared for histopathological procedures is summarized in Table 11.

Selected References

Kunwar, I. K., Singh, T., and Sinclair, J. B. 1985. Histopathology of mixed infections by *Colletotrichum truncatum* and *Phomopsis* spp. or *Cercospora sojina* in soybean seeds. Phytopathology 75:489-492.

Singh, T., and Sinclair, J. B. 1985. Histopathology of *Cercospora sojina* in soybean seeds. Phytopathology 75:185-189.

Yorinori, J. T. 1980. *Cercospora sojina*: Pathogenicity, new races, and seed transmission in soybeans. Ph.D. thesis, University of Illinois at Urbana-Champaign. 174 pp.

(Prepared by M. M. Kulik and J. B. Sinclair)

Chaetomium cupreum

Chaetomium spp. and *C. cupreum* Ames have been reported to be associated with soybean seeds in Brazil and the United States. *C. cupreum* produces a compound that inhibits seed germination and the growth of other fungi associated with soybean seeds. It also produces a copper-colored pigment on agar, which can be confused with that produced by *Cercospora kikuchii*.

Selected References

Jordan, E. G., Manandhar, J. B., Thapliyal, P. N., and Sinclair, J. B. 1988. Soybean seed quality of 16 cultivars and four maturity groups in Illinois. Plant Dis. 72:64-67.

Yeh, C. C., and Sinclair, J. B. 1980. Effect of *Chaetomium cupreum* on seed germination and antagonism to other seedborne fungi of soybean. Plant Dis. 64:468-470.

(Prepared by M. M. Kulik and J. B. Sinclair)

Colletotrichum truncatum

Seeds colonized by *Colletotrichum truncatum* (Schw.) Andrus & W. D. Moore may show no symptoms or may develop brown staining (Fig. 10J and Plate 5) or small, irregular gray areas with black specks. The fungus is confined at first to the seed coat (Fig. 87), as is *Phomopsis longicolla* Hobbs (see *Diaporthe-Phomopsis* Complex). Infected seeds may die during germination or, if they do germinate, may produce infected seedlings. Pods and stems infected by the fungus do not necessarily show symptoms (see Anthracnose).

The appearance of *C. truncatum* in soybean seed coats prepared for histopathological procedures is summarized in Table 11.

C. truncatum and *P. longicolla* have an additive effect on the destruction of seed coat tissues when both are present.

Selected References

Hepperly, P. R., Mignucci, J. S., Sinclair, J. B., and Mendoza, J. B. 1983. Soybean anthracnose and its seed assay in Puerto Rico. Seed Sci. Technol. 11:371-380.

Khare, M. N., and Chacko, S. 1983. Factors affecting seed infection and transmission of *Colletotrichum dematium* f. sp. *truncata* in soybean. Seed Sci. Technol. 11:853-858.

Kunwar, I. K., Singh, T., and Sinclair, J. B. 1985. Histopathology of mixed infections by *Colletotrichum truncatum* and *Phomopsis* spp. or *Cercospora sojina* in soybean seeds. Phytopathology 75:489-492.

(Prepared by M. M. Kulik and J. B. Sinclair)

Diaporthe-Phomopsis Complex

The fungus most frequently isolated from seeds infected with members of the *Diaporthe-Phomopsis* complex (Plate 10) is *P. longicolla* Hobbs (syns. *P. phaseoli* (Desm.) Sacc. and *P. glycines* Petrak). Others are *D. phaseolorum* (Cke. & Ell.) Sacc. var. *sojae* (Lehman) Wehm. and *D. phaseolorum* var. *caulivora* Athow & Caldwell. The appearance of *Phomopsis* spp. in soybean seed coats prepared for histopathological procedures is summarized in Table 11.

The significance of seeds as a source of inoculum for outbreaks of pod and stem blight and stem canker is not known, but infected seeds can introduce the pathogens into uninfested areas (see Pod and Stem Blight and Phomopsis Seed Decay; Stem Canker; Phomopsis Seed Decay).

Selected Reference

Anderson, T. R. 1985. Seed molds of soybean in Ontario and the influence of production area on the incidence of *Diaporthe phaseolorum* var. *caulivora* and *Phomopsis* sp. Can. J. Plant Pathol. 7:74-78.

(Prepared by M. M. Kulik and J. B. Sinclair)

Fusarium spp.

The following *Fusarium* spp. have been isolated from soybean seeds: *F. dimerum* Penz., *F. equiseti* (Corda) Sacc., *F. graminearum* Schwabe (teleomorph *Gibberella zeae* (Schw.) Petch), *F. moniliforme* Sheld. (teleomorph *G. fujikuroi* (Sawada) Ito), *F. oxysporum* Schlecht. ex Fr., *F. rigidiusculum* (Brick) Snyd. & Hans. (teleomorph *Calonectria rigidiuscula* (Berk. & Br.) Sacc.), *F. semitectum* Berk. & Rav., and *F. solani* (App. & Wollenw.) Snyd. & Hans.

The first reported occurrence of preharvest infection of soybean seeds by *F. graminearum* was recorded in the midwestern United States in 1986. A large portion of seeds became infected in September, just before harvest, following an unusually long period of warm, wet weather. Infected seeds were stained pink to red. Similarly discolored seeds were found in 1987, but to a lesser extent.

The appearance of *Fusarium* spp. in soybean seed coats prepared for histopathological procedures is summarized in Table 11.

Selected References

Dhingra, O. D., Sediyama, T., and Sediyama, F. 1979. Effect of planting and harvest time on seed infection of soybean by *Phomopsis sojae* and *Fusarium semitectum*. Fitopatol. Bras. 4:467-472.

Hepperly, P. R. 1985. *Fusarium* species and their association with soybean seed under humid tropical conditions in Puerto Rico. J. Agric. Univ. P.R. 69:25-33.

Jordan, E. G., Manandhar, J. B., Thapliyal, P. N., and Sinclair, J. B. 1988. Soybean seed quality of 16 cultivars and four maturity groups in Illinois. Plant Dis. 72:64-67.

Wicklow, D. T., Bennett, G. A., and Shotwell, O. L. 1987. Secondary invasion of soybeans by *Fusarium graminearum* and resulting mycotoxin contamination. Plant Dis. 71:1146.

(Prepared by M. M. Kulik and J. B. Sinclair)

Macrophomina phaseolina

Soybean seeds infected by *Macrophomina phaseolina* (Tassi) Goid., the cause of charcoal rot (see Charcoal Rot), may be asymptomatic or show indefinite black spots or blemishes on the seed coat. Sometimes microsclerotia are produced in fissures and cracks in the seed coats of infected seeds (Fig. 88). Fungal hyphae penetrate inter- and intracellularly in seed coat tissues and form microsclerotia in asymptomatic seeds when conditions are favorable for germination. The appearance of *M. phaseolina* in soybean seed coats prepared for histopathological procedures is summarized in Table 11.

Seedlings from infected seeds can become infected, with or without symptoms; infected seedlings are unthrifty or affected with postemergence damping-off.

Fig. 87. Cross section of a soybean seed coat infected with *Colletotrichum truncatum*, in a bright-field photomicrograph showing a mature acervulus with setae and hypha (arrow) from initial colonization of the aleurone layer. (Courtesy A. Rodriguez-Marcano)

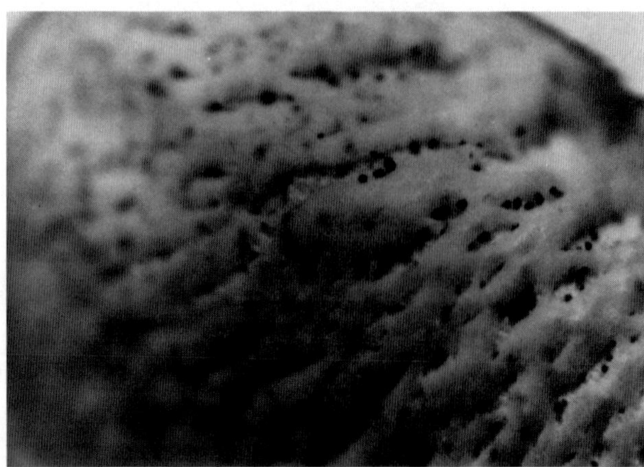

Fig. 88. Soybean seed naturally infected with *Macrophomina phaseolina*, with black microsclerotia in the fissured seed coat. (Courtesy I. K. Kunwar)

Selected Reference

Kunwar, I. K., Singh, T., Machado, C. C., and Sinclair, J. B. 1986. Histopathology of soybean seed and seedling infection by *Macrophomina phaseolina*. Phytopathology 76:532-535.

(Prepared by M. M. Kulik and J. B. Sinclair)

Yeast Spot (Nematospora Spot)

Yeast spot, or Nematospora spot, was first observed in soybeans in the United States in 1928. It was reported as a serious disease affecting soybean seeds in North Carolina and Oklahoma in 1943 and in Kansas and Missouri in 1964. Since then it has been reported in soybean-growing areas throughout the world.

Diseased seeds are discolored and have a lowered oil content. Badly damaged seeds are downgraded in commercial trade. Seeds infected early are small, shriveled, and lightweight. The yeast also infects at least 18 other hosts.

Symptoms

Yeast spot is primarily a disease of soybean seeds, although pods often show small discolored areas where they contact infected seeds. When infection occurs during pod formation, seeds fail to develop and pods drop prematurely.

Diseased tissue of young seeds is pale yellow to grayish brown and slightly sunken. The sunken spots are light or cream-colored on more mature seeds. Affected areas of cotyledons are off-white and cheesy in texture (Fig. 89). Varying amounts of dead tissue may extend into infected embryos. The seed coat remains intact except for minute, discolored punctures, which may be seen upon close examination. Severely infected seeds do not mature and are greatly shrunken and wrinkled.

Causal Organism

Nematospora coryli Pegl. (syn. *N. phaseoli* Wingard) causes yeast spot. *N. lycopersici* Schneider and *Ashby gossypii* (Ashby & Nowell) Guilliermond (syns. *N. gossypii* Ashby & Nowell and *Ashbya gossypii* (Ashby & Nowell) Ciferri & Fragoso) have also been reported on soybean seeds. Soybean seeds infected with *N. lycopersici* were reported in Kansas and Missouri in 1964, and seeds infected with *A. gossypii* were reported in Missouri in 1970.

The vegetative growth of *N. coryli* may consist of individual cells or septate strands resembling mycelium, 90–140 × 2.3–3.5 μm. Buds arise at the cross-walls of the strand. Individual cells are elliptical (8–14 × 6–10 μm), spherical (12–20 μm in diameter), or, occasionally, slender and tapering and shaped like a tennis racket.

Asci and ascospores are formed on seeds and in nutrient broth. The asci are cylindrical (60–85 × 10–12 μm), with broadly rounded ends, and contain eight ascospores in two groups of four. The ascospores (40–60 × 2.5–3.5 μm) are slender, two-celled, and slightly constricted at the septum, with a sharp tip; the base extends into a slender, nonmotile whip about 1.25 times as long as the spore.

Ascospores germinate by forming a germ tube or by budding. Usually, the basal cell swells at the septum, forming a sphere about 6 μm in diameter, from which a germ tube protrudes. The germ tube develops into a septate, branched, mycelium-like strand. Less commonly, round to elliptical cells bud from the spherical basal cell.

N. coryli grows well at 30–40°C in nutrient broth supplemented with 1% glucose, 0.5% yeast extract, and 1,000 μg of streptomycin sulfate per milliliter. Maximum growth is reached in 24 hr at 40°C. At 18°C, the yeast grows poorly, and asci and ascospores develop slowly. Buffered media of pH 6 are optimal for growth. The yeast lowers the pH of nonbuffered media to a level at which growth is inhibited.

Transmission

N. coryli is usually transmitted by the green stinkbug, *Acrosternum hilare* (Say) (Fig. 90). Other species capable of transmitting the yeast to soybeans are *Euschistus euschistoides* (Vollenhoven), *E. servus* (Say), *E. tristigmus* (Say), *E. variolarius* (Palisot de Beauvois), and *Thyanta custator* (Fabricius).

For seeds to become infected, yeast-infested stinkbugs must pierce the pod wall and feed on the developing bean, forcing yeast cells into it. Only yeast cells near the mouthparts are transmitted. Nymphs cannot transmit the yeast after molting. Nymphs and newly emerged adults must feed on a source of *N. coryli* after each molt to become transmitters. The yeast survives in adults throughout the insect's life.

Little is known about the epidemiology and other aspects of the disease.

Control

1. Plant seeds relatively free of the pathogen.
2. Control stinkbugs with insecticides when the average population is one nymph per 30 cm of row.

Selected References

Batra, L. R. 1963. Nematosporaceae (Hemiascomycetidae): Taxonomy, pathogenicity, distribution, and vector relations. U.S. Dep. Agric., Tech. Bull. 1469. 71 pp.

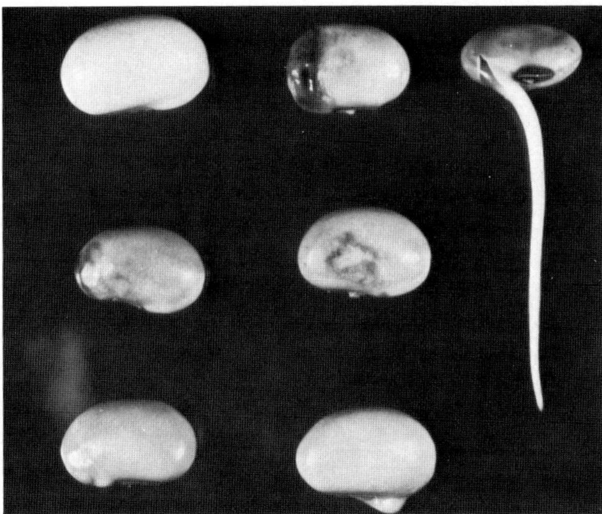

Fig. 89. Soybean seeds with whitish or cheesy spots, symptoms of yeast spot, caused by *Nematospora coryli*, and an asymptomatic seed (upper left). (Courtesy J. T. Yorinori)

Fig. 90. Green stinkbug, *Acrosternum hilare*, which transmits the yeast spot fungus, *Nematospora coryli*. (Courtesy C. A. Thomas)

Clarke, R. G., and Wilde, G. E. 1971. Association of the green stinkbug
and the yeast spot disease organism of soybeans. III. Effect on
soybean quality. J. Econ. Entomol. 64:222-223.
Daugherty, D. M. 1967. Pentatomidae as vectors of yeast-spot disease
of soybean. J. Econ. Entomol. 60:147-152.

(Prepared by M. M. Kulik and J. B. Sinclair)

Peronospora manshurica

Peronospora manshurica (Naum.) Syd. ex Gäum. forms a
milky white encrustation of mycelium and oospores on seeds
(Fig. 17). Seeds of highly susceptible cultivars may be smaller or
lighter in weight than normal seeds. Occasionally mycelium
may be found in cell layers of the seed coat. Infected seeds may
not emerge or may produce systemically infected seedlings. (See
Downy Mildew.)

Selected Reference

Inaba, T. 1985. Seed transmission of downy mildews of spinach and
soybean. (In Japanese) JARQ 19:26-31.

(Prepared by M. M. Kulik and J. B. Sinclair)

Phomopsis Seed Decay

Phomopsis seed decay is caused primarily by *Phomopsis
longicolla* Hobbs, which is part of the *Diaporthe-Phomopsis*
complex (see Pod and Stem Blight and Phomopsis Seed Decay;
Stem Canker). *P. longicolla* is frequently recovered from
infected seeds more often than *D. phaseolorum* (Cke. & Ell.)
Sacc. var. *sojae* (Lehman) Wehm. (anamorph *P. sojae*
Lehman), which is recovered more often than *D. phaseolorum*
var. *caulivora* Athow & Caldwell.

Severely infected seeds are shriveled, elongated, and cracked
and appear white and chalky (Fig. 46 and Plate 10). They
usually do not germinate or are delayed in germination. Seeds
may also be infected and not show symptoms. Seed infection
causes pre- and postemergence damping-off, and under severe
conditions stands can be reduced to the point of reducing yields.

The fungi at first colonize seed coats and then cotyledons and
plumules (Fig. 91). The appearance of *Phomopsis* spp. in
soybean seed coats prepared for histopathological procedures is
summarized in Table 11.

Infected seeds are a minor source of primary inoculum,
compared with infested crop debris and soil. Soybean pods can
become infected at any time after they are formed, but
significant seed infections do not occur before physiological
maturity. Seed infection tends to be more severe when the
weather delays harvest, when early-maturing cultivars are
grown, or when crops are grown in regions where warm, humid
weather prevails at harvesttime.

Selected References

Kulik, M. M., and Schoen, J. F. 1981. Effect of seedborne *Diaporthe
phaseolorum* var. *sojae* on germination, emergence, and vigor of
soybean seedlings. Phytopathology 71:544-547.
Kunwar, I. K., Singh, T., and Sinclair, J. B. 1985. Histopathology of
mixed infections by *Colletotrichum truncatum* and *Phomopsis* spp.
or *Cercospora sojina* in soybean seeds. Phytopathology 75:489-492.
Singh, T., and Sinclair, J. B. 1986. Further studies on the colonization
of soybean seeds by *Cercospora kikuchii* and *Phomopsis* sp. Seed
Sci. Technol. 14:71-77.
Spilker, D. A., Schmitthenner, A. F., and Ellett, C. W. 1981. Effects of
humidity, temperature, fertility, and cultivar on the reduction of
soybean seed quality by *Phomopsis* sp. Phytopathology
71:1027-1029.
Thomison, P. R. 1985. Factors affecting the severity of Phomopsis seed

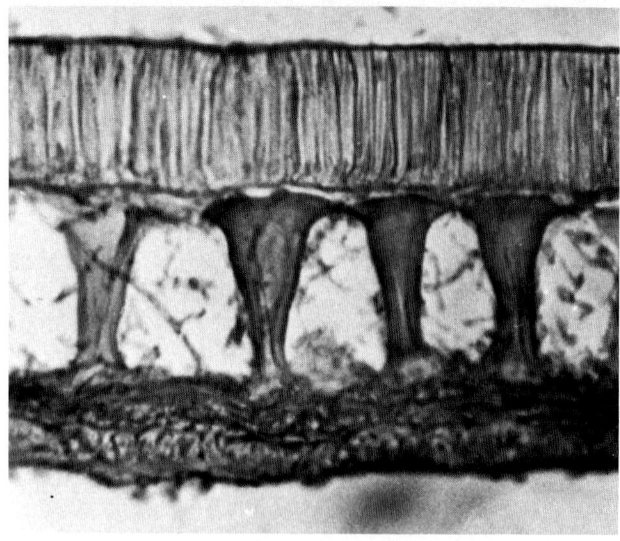

Fig. 91. Cross section of a soybean seed coat with hyphae of
Phomopsis longicolla colonizing the hourglass cell layer and
penetrating the upper palisade cell layer and the lower aleurone
cell layer, in a bright-field photomicrograph (×300). (Courtesy
M. B. Ilyas)

decay in soybeans. Pages 495-502 in: Proc. World Soybean Res.
Conf. III. R. Shibles, ed. Westview Press, Boulder, CO.

(Prepared by M. M. Kulik and J. B. Sinclair)

Other Seedborne Fungi

Numerous other genera of fungi have been reported to be
seedborne in soybeans, including *Ascochyta, Cladosporium,
Corynespora, Curvularia, Drechslera, Epicoccum, Helminthosporium, Macrophoma, Monilia, Nigrospora, Paecilomyces,
Pestalotia, Phialophora, Phoma, Phyllosticta, Phytophthora,
Rhizoctonia, Rhizopus, Sclerotium, Septoria, Stemphylium,
Thielaviopsis*, and *Verticillium*. Some of them are recognized as
pathogens of soybeans, but no documented evidence suggests
that they affect seed quality or that seed infection by these fungi
results in disease development.

Selected Reference

Malone, J. P., and Muskett, A. E. 1964. Seed-borne fungi. Proc. Int.
Seed Test. Assoc. 29:179-384.

(Prepared by M. M. Kulik and J. B. Sinclair)

Postharvest Pathology

Deterioration of soybeans by fungi in the field following
weather-delayed harvests and in storage is a serious economic
problem for both producers and processors. Deterioration
caused by fungi results in increased damage discounts against
the producer, reduced oil and meal quality, and reduced
storability of soybean products. In some situations, mycotoxins
produced by *Aspergillus flavus* Lk. ex Fr., *A. parasiticus*
Speare, and *Fusarium graminearum* Schwabe can cause
problems for animals consuming soybeans. Losses are sporadic
and difficult to assess but are estimated to exceed $20 million
annually in the United States. This estimate does not include
reductions in prices due to damage discounts assessed when
mold-damaged soybeans are marketed. Price reductions are
frequently in the 30–50% range and may approach $100 million
annually in the United States.

Field Fungi

Fungi causing deterioration in the field following weather-delayed harvests include species of *Phomopsis, Alternaria, Chaetomium,* and *Fusarium,* as well as others that infect pods and seeds. These fungi, described elsewhere in this compendium, are the cause of "weather damage." They generally cease activity when seed moisture drops below 18–20%, and they begin to die under prolonged storage at low moistures. However, damage from field fungi increases the risk of damage from storage fungi, such as species of *Aspergillus* and *Penicillium,* which can grow in soybeans if the average seed moisture content is kept above 12%.

Storage Fungi

Serious economic losses frequently occur in storage as a result of deterioration caused by several species of *Aspergillus* (Fig. 92), *Penicillium, Paecilomyces, Scopulariopsis, Rhizopus, Mucor,* and *Fusarium* and by yeasts and bacteria. Except for *Aspergillus* and *Penicillium,* these pathogens are associated only with advanced decay. More than 150 species have been associated with stored seeds. These microorganisms produce both heat and water from their respiration and are capable of raising temperatures to levels higher than 76° C. As a result of exothermic oxidative chemical reactions, heating to ignition can occur when sufficient oxygen is available. At temperatures below the heat of ignition, soybeans can appear to be charred and blackened, in a condition known as bin burning (Fig. 93). Bin-burned soybeans contain excessive amounts of free fatty acids, which cause off-flavors in soybean oil or oil products and dramatically reduce their shelf life.

Storage fungi in the genera *Aspergillus* and *Penicillium* are especially adapted for growth in soybeans with moisture contents above 12.0–13.0% (i.e., in equilibrium with relative humidities of 70–75%). At moisture contents higher than 20%, these fungi are commonly succeeded by species of *Mucor, Rhizopus,* and *Fusarium* and by yeasts and thermophilic bacteria. Each fungus has a distinct optimal environment for growth, and the presence or absence of a given fungus can provide evidence of the moisture, temperature, and degree of deterioration of a given soybean lot. Information on the temperature and moisture ranges of *Aspergillus* spp. and *Penicillium* spp. commonly found in damaged soybeans is presented in Table 12.

The respiration of fungi increases the temperature and available moisture in the grain mass and allows for a more diverse fungal community growing under increasingly more favorable conditions. Generally, the growth of *A. restrictus* G. Smith produces little detectable heating; however, as moisture increases, *A. glaucus* Lk. ex Fr. becomes active, producing temperature increases of 10–15° C and moisture, particularly as temperature and moisture reach the optimal range for that species. At moisture contents above 14.0–15.0%, the growth of *A. candidus* Lk. and *A. flavus* rapidly raises the temperature to 55° C. At this point in the process of deterioration, hot spots are usually detectable in the grain mass, and soybeans become clumped and turn dark brown. Seeds heated to ignition are blackened, and cavities are formed in the cotyledons, as observed in cross section (cavities are not present in the cotyledons of seeds heated until they turn brown). Further, blackened seeds lose little or no weight when heated to 220° C for 3–6 hr.

Losses to storage mold fungi can be avoided. Storage molds are controlled primarily by managing grain moisture or

Fig. 92. Growth of the storage fungi *Aspergillus restrictus* (left) and *A. halophilicus* (right) on soybean seeds stored at 12.5–13.0% moisture for 30 days at room temperature. (Courtesy C. M. Christensen)

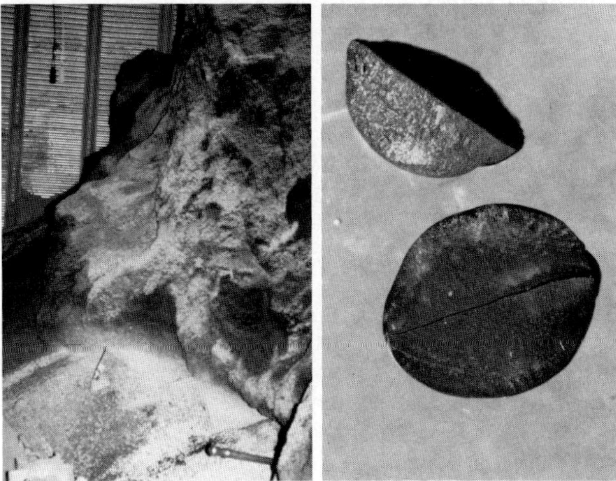

Fig. 93. Soybeans in flat storage with bin burning associated with high-moisture weed seeds in the spout line (left); bin-burned seeds (right). (Courtesy B. J. Jacobsen [left] and R. A. Meronuck [right])

Table 12. Temperatures, Relative Humidities, and Equilibrium Moisture Contents for Species of *Aspergillus* and *Penicillium* Commonly Associated with Decay of Soybeans in Storage[a]

Fungus	Temperature (°C)[b]			Relative Humidity (%)		Minimum Equilibrium Moisture Content (%)[c]
	Min.	Opt.	Max.	Min.	Opt.	
Aspergillus candidus	—	32	35	75	80	14.0
A. flavus	15	32	55	81	>85	15.0
A. glaucus group	2	23	35	70	75	13.0
A. halophilicus	—	25	35	65	75	12.0
A. niger	8	35	46	77	—	14.0
A. ochraceus	—	25	42	80	95	14.0
A. parasiticus	15	32	55	81	>85	15.0
A. restrictus	4	25	35	65	75	12.0
A. terreus	—	30	45	80	90	14.0
Penicillium spp.	4	25	35	85	90	15.0–17.0

[a] Data from Christensen and Sauer, 1982.
[b] Min. = minimum; Opt. = optimum; Max. = maximum.
[c] Figures are approximations, and actual figures may vary by 10%.

temperature so that the storage environment is unfavorable for mold growth. Soybeans with a moisture content of less than 12% in storage (in equilibrium with a relative humidity less than 65%) do not support the growth of storage molds. Temperatures below 10°C also suppress mold growth. The effect of temperature is illustrated by the fact that soybeans can be stored more than 1 year at 14.0–14.5% moisture if the temperature is kept at 6–8°C, whereas at 30°C nearly 100% of the seeds become mold-damaged within 3–5 weeks.

Storage losses occur when moisture is present in amounts sufficient to allow mold growth. Since mold respiration both raises the temperature and increases the moisture in storage, serious losses can occur even if the average grain moisture content is less than 12%, for several reasons. First, in some volume of seeds in storage, the moisture content exceeds 12%, and fungi growing on these seeds produce heat and moisture, which allow more mold growth on adjacent seeds. Second, weed seeds harvested with the soybeans have moisture contents high enough to support fungal growth. Weed seeds frequently congregate in cone-shaped compacted layers that accumulate in the grain mass under the loading spout. These layers of fine material, including broken soybeans as well as weed seeds, are known as spout lines. Hot spots often develop within spout lines. Third, ice harvested with soybeans eventually becomes free water. Fourth, moisture transfer from the air can occur when the air and the grain are at different temperatures; the unequal temperatures cause convection currents, resulting in condensation in cool areas. This phenomenon often occurs in the fall, when warm grain is placed in a storage bin and the air becomes cooler, or, more frequently, when the grain mass is cold following winter storage in temperate climates and is then exposed to warm, moist air during unloading and transfer. Damage resulting from moisture transfer is frequently found in the upper layers near the center of the bin.

Control of Storage Fungi

1. The highest seed moisture should not exceed 12%. If temperatures are less than 10°C, storage at 13–14% moisture is safe for up to 6 months. Moisture-measuring equipment should be regularly calibrated. Storability is determined by the wettest seeds, not the average moisture. High-moisture seeds adjacent to low-moisture seeds may lose moisture over time, but the average moisture of a grain lot does not provide enough information to evaluate potential storage mold activity.

2. Clean soybean lots to remove broken seeds and weed seeds. Cleaning not only removes potentially high-moisture weed seeds but helps to eliminate spout lines, which slow air movement in the storage bin. Aeration by means of fans in the sizes common in most storage structures can be used to regulate temperatures. Larger airflows are necessary for affecting equilibrium relative humidities in the grain mass.

3. Monitor temperatures. Where possible, the temperature of the grain mass should be maintained within 5–10°C of the average monthly outdoor temperature. The grain mass temperature should be constantly monitored with preplaced thermocouple cables or temperature probes. Any sign of temperature increase should signal advanced storage decay, and immediate action should be taken to cool the grain and move it before further losses occur. A rise of more than 1–3°C indicates that advanced spoilage is under way. Grain is an excellent insulator, and temperatures of 40–50°C may be undetectable as little as 0.3–1.0 m away.

4. Control insects associated with stored grain. Insect activity can raise both grain temperature and moisture.

Selected References

Christensen, C. M., ed. 1982. Storage of Cereal Grains and Their Products, 3rd ed. American Association of Cereal Chemists, St. Paul, MN. 544 pp.
Christensen, C. M., and Dorworth, C. E. 1966. Influence of moisture content, temperature, and time on invasion of soybeans by storage fungi. Phytopathology 56:412-418.
Christensen, C. M., and Meronuck, R. A. 1986. Quality Maintenance in Stored Grains and Seeds. University of Minnesota Press, Minneapolis. 138 pp.
Christensen, C. M., Meronuck, R. A., Steele, J. A., and Behrens, J. C. 1973. Some morphological and chemical characteristics of binburned and fireburned soybeans. Trans. ASAE 18:899-901.

(Prepared by B. J. Jacobsen and R. A. Meronuck)

Seedborne Viruses

Of all the viruses that can infect soybeans, soybean mosaic virus (SMV) causes the most conspicuous symptoms on seeds (see Soybean Mosaic Virus). Seeds from SMV-infected plants may or may not show symptoms and may be smaller than seeds from healthy plants (Fig. 77 and Plates 18 and 19). Yellow-seeded cultivars infected with SMV may produce seeds that are mottled brown or black, depending on the hilum color (Fig. 77). The mottle usually appears as a "bleeding" of the hilum pigment over the seed coat, a condition often referred to as bicolor. The mottling may also appear on seeds from plants with the gene that inhibits the expression of hilum color.

Selected References

Bowers, G. R., Jr., and Goodman, R. M. 1982. New sources of resistance to seed transmission of soybean mosaic virus in soybeans. Crop Sci. 22:155-156.
Goodman, R. M., and Oard, J. H. 1980. Seed transmission and yield losses in tropical soybeans infected by soybean mosaic virus. Plant Dis. 64:913-914.
Suteri, B. D. 1981. Effect of soybean mosaic virus on seed germination and its transmission through seeds. Indian Phytopathol. 34:370-371.

(Prepared by M. M. Kulik and J. B. Sinclair)

Other Pathogens Associated with Seeds

The resting stages of several fungi are found as contaminants in soybean seed lots. For example, the galls of an unidentified oomycete, carried in small soil balls referred to as soil peds, have been found among soybean seeds (Fig. 94). The large, black sclerotia of *Sclerotinia sclerotiorum* (Lib.) d By. (Fig. 55) are not easily separated from seed lots during harvest or routine cleaning. Smut galls of *Melanopsichium pennsylvanicum* Hirsch. from *Polygonum pennsylvanicum* L. (smartweed), referred to in the literature as soybean smut, have also been reported in seed lots in Australia and the United States. The galls look like soil peds (Fig. 95). Smut galls and seeds of the smartweed host may be carried in soybean seed lots, and both

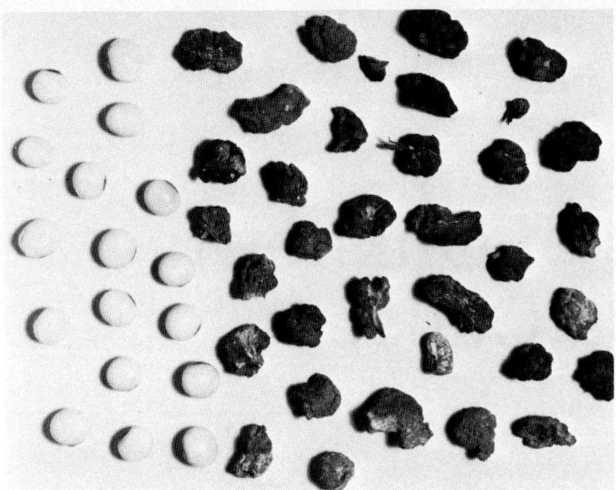

Fig. 94. Galls of an unidentified oomycete (right) found in soil peds associated with soybean seed lots, compared with soybean seeds. (Courtesy J. W. Gerdemann)

Fig. 95. Galls of *Melanopsichium pennsylvanicum* (left and right) from *Polygonum pennsylvanicum* (once referred to as soybean smut), compared with soybean seeds. (Courtesy C. T. Schiller)

the fungus and the weed could become established in new areas in this manner.

In addition, soil peds have been associated with cysts of the soybean cyst nematode (Plate 28) (see Soybean Cyst Nematode) and sclerotia of *S. sclerotiorum* (see Sclerotinia Blight) and may be mistaken for them unless examined carefully. The cysts of the soybean cyst nematode remain alive for up to 15 months in soil peds taken from seed samples.

Selected References

Gerdemann, J. W., and Chamberlain, D. W. 1969. Galls containing oospores of an unidentified fungus found as a contaminant of soybean seed. Plant Dis. Rep. 53:978-979.

Langdon, R. F. N., and Cusack, A. N. 1978. Soybean smut. Australas. Plant Pathol. 7:43-44.

Schiller, C. T., and Sinclair, J. B. 1979. Soybean seed lot contamination by *Melanopsichium pennsylvanicum* smut galls. Phytopathology 69:605-608.

(Prepared by M. M. Kulik and J. B. Sinclair)

Part II. Diseases of Unknown or Uncertain Cause

Several soybean diseases have been observed for which no cause has been determined. For example, sudden death syndrome has been observed for many years in the United States, soybean plants with severe phyllody have been reported in Nigeria and Tanzania, and plants with flower abortion and defoliation have been reported in Peru (Fig. 96). Sudden death syndrome and two leaf disorders are described in this section.

Foliage Blight

A foliage blight is prevalent in the high hills of Nepal (at elevations around 2,000 m). No cause is known. Soybeans are predisposed to the blight during monsoon rains and cool weather. The blight may be severe during prolonged rainy periods. No comprehensive description of the disorder has been published.

Symptoms appear first on the lower leaves and advance to the upper ones. Discrete, water-soaked, orange-brown specks form on leaves. Later, these enlarge to round or irregularly shaped orange-brown lesions, 2–3 cm in diameter, with definite, dark brown margins (Fig. 97). The spots may continue to enlarge, and several may coalesce if cool, misty weather persists. Elongated spots occasionally develop on petioles, stems, and pods. Severely affected leaves drop off.

(Prepared by J. B. Manandhar)

Sudden Death Syndrome

Sudden death syndrome (SDS) is a mid- to late-season, apparently soilborne disease that occurs primarily in soybean fields with high yield potential. It was first observed in Arkansas in 1971 by H. J. Walters and was first described in 1983. It has since appeared in Illinois (1985), Indiana (1985), Kentucky (1984), Mississippi (1984), Missouri (1984), and Tennessee (1984). SDS can result in minor or almost total yield losses, depending on when it develops, in portions or all of severely affected fields. Yield losses are the result of flower and pod abortion, pod drop, and reduced seed size.

Symptoms

Initial symptom development is associated with soybeans in the early reproductive growth stages and varies geographically. In the southern portions of its range, SDS commonly appears as early as 2 to 3 weeks before flowering, with extensive symptom development after flowering. In the northern part of its range, where earlier-maturing cultivars are grown, symptoms rarely appear before the middle of the pod-fill stage.

Symptoms first appear on leaves as scattered, interveinal, chlorotic spots or blotches (Plates 32 and 33). The chlorotic areas may become necrotic or may enlarge and coalesce,

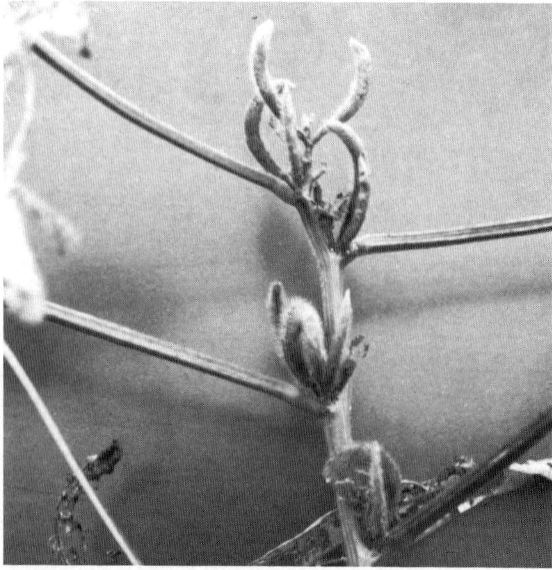

Fig. 96. Soybean plants from Peru with symptoms of a disease of uncertain cause: symptoms on the upper portion of the plant (above); sterile, erect pods (below). (Courtesy C. R. Valles [above] and M. E. Irwin [below])

forming interveinal chlorotic streaks. These streaks eventually become necrotic, and only the midvein and major lateral veins remain green. Severely affected leaflets either drop off, leaving the petioles attached, or curl upward and remain attached to the plant (Fig. 98). Severe foliar symptoms give affected areas in the field a tan to brown cast and may be the first evidence of the disease. The effect of SDS on yield depends on the growth stage at the time of initial symptom development and the rapidity and severity of foliar symptom development. Severe symptom development in the early reproductive stages may lead to flower and pod abortion and pod drop; severe symptom development during pod fill may result in reduced seed size.

Root symptoms precede foliar symptoms and are characterized by deterioration of taproots, lateral roots, and nitrogen-fixing nodules. The cortex of affected taproots is a light gray-brown. The discoloration extends up the stem several nodes in the vascular tissue, but the pith remains white.

Foliar symptoms of stem canker and brown stem rot are similar to those of SDS. Stem canker differs from SDS by the presence of a canker on the lower stem. Brown stem rot is distinguished from SDS by a distinct discoloration of the pith.

Possible Causal Agents

The cause of SDS remains uncertain, but two possible causal agents have been proposed: *Xanthomonas campestris* (Pam.) Dowson and *Fusarium solani* (App. & Wollenw.) Snyd. & Hans. Both microorganisms have caused SDS-like symptoms in greenhouse and growth chamber tests. *X. campestris* has been isolated from the vascular tissue of plants with SDS. The bacterium is characterized by gram-negative rods that produce small, yellow, glistening, raised colonies with an entire border on nutrient agar. *F. solani* has been isolated from the roots and vascular tissue of plants with SDS and also from the soil of affected areas in the field. The fungus is characterized by slow, appressed growth on potato-dextrose agar, often staining the agar a dark maroon-brown. Abundant macroconidia (30–65 × 6–8 μm) having three to five septa are produced in a slimy, blue mass with few microconidia, if any.

Epidemiology

SDS is prevalent during cool, wet growing seasons. Hot, dry weather appears to slow or arrest the disease, although it has been severe under these conditions. Irrigated soybeans are more prone to the disease than nonirrigated soybeans. SDS may occur in spots in a field or pervade an entire field. Fields in which the disease is present are likely to develop it in subsequent years. Cropping sequence may have little impact on SDS, since severe outbreaks have occurred in fields previously cropped with cotton, maize, rice, sorghum, or soybeans. Moderate to high populations of the soybean cyst nematode (*Heterodera glycines* Ichinohe) are usually, but not always, present in fields with SDS. Generally the disease occurs in fields where soybean-growing conditions are optimal and yield potentials are high.

Control

1. Plant cultivars with a history of light reaction to SDS.

2. Plant an infested field with cultivars of different maturity groups, or plant portions of the field at different times.

3. Cultural practices to reduce plant stress, control of the soybean cyst nematode, and improvements in drainage may help plants tolerate SDS.

Fig. 97. Lesion of uncertain cause (lower left corner) on a soybean leaflet grown in Nepal. (Courtesy J. B. Manandhar)

Fig. 98. Soybean plant with advanced symptoms of sudden death syndrome, a disease of unknown cause: defoliation, with petioles remaining attached. (Courtesy J. C. Rupe)

Selected References

Hirrel, M. C. 1983. Sudden death syndrome of soybean—A disease of unknown etiology. (Abstr.) Phytopathology 73:501-502.

Rupe, J. C., and Weidemann, G. J. 1986. Pathogenicity of a *Fusarium* sp. isolated from soybean plants with sudden death syndrome. (Abstr.) Phytopathology 76:1080.

Yopp, J., Krishnamani, R. S., Bozzola, J., Richardson, J., Myers, O., and Klubek, B. 1986. Presumptive role of a pathovar of *Xanthomonas* in sudden death syndrome of soybean. Microbios Lett. 32:75-79.

(Prepared by J. C. Rupe, M. C. Hirrel,
and D. E. Hershman)

Yellow Leaf Spot

Yellow leaf spot has been reported only in Thailand. It was found in the Chiang Mai area, where 70–80% of plants were affected, and in the central and western areas of Khon Kaen, where 20–30% of plants were affected. No cause is known.

The disease causes rapid defoliation, poor pod fill, and reduced seed size. Lesions develop from pinpoint spots similar to those of bacterial pustule (see Bacterial Pustule). Later, yellow halos, larger than those of bacterial pustule, develop around the enlarging spots (Fig. 99). Young plants may have more spots than older ones, and spots may be reddish brown. After 1 to 2 weeks, infected leaves dry up and drop off. The disease spreads rapidly during heavy rainstorms and under hot, humid conditions. The most susceptible cultivars are SJ-1, SJ-2, and SJ-4.

Selected Reference

Poonpolgul, S., and Pupipat, U. 1978. Soybean diseases in Thailand. (In Thai; English summary) Kasesart J. 12:143-154.

Fig. 99. Symptoms of yellow leaf, a disease of uncertain cause widespread in soybean-growing areas of Thailand. (Courtesy S. P. Poonpolgul)

Part III.
Noninfectious or Stress Diseases

Noninfectious or stress "diseases" are due to causes other than infectious agents, such as bacteria, fungi, nematodes, or viruses. Noninfectious disease factors include excesses, deficiencies, or imbalances of soil nutrients or water; extremes in soil acidity or alkalinity; extremes in temperature; air pollutants; excesses of pesticides; and mechanical or other injury. The severity and type of injury vary with the plant growth stage, the time of the disturbance, and the plant part involved. Symptoms of noninfectious agents are often confused with symptoms caused by infectious agents. Stress conditions can predispose soybean plants to attack by infectious agents.

Crusting and Compaction

When heavy rain falls on fine-textured soil, a hard crust may form and prevent germinated soybean seeds from emerging (Fig. 100 and Plate 34). The stems of such seedlings become thickened, and in severe cases, the hypocotyl arch is cracked or broken. Rotary hoeing or hand-hoeing to break up the surface crust may aid emergence.

Frost

Frost-damaged plants are common in temperate growing areas. Often only the upper part of the plant is killed or damaged (Plates 35 and 36), and regrowth soon appears at one of the unaffected nodes. If the new growth is strong and vigorous, and if few plants are damaged severely, replanting may not be needed.

Hail

Leaves injured by hail are torn and ragged; large areas of tissues may be beaten away. Stems may be cut off or broken or may have sunken, dark areas (Plate 37). Pathogenic fungi may colonize hail-damaged areas and produce stem-girdling cankers or weaken the stems. Most damage is usually on the side of the plant facing the prevailing winds of the storm. The decision to replant depends upon the extent of damage, the stage of plant development, and the regrowth that appears. In general, yield is reduced in direct proportion to the percentage of the leaf surface removed and the number of plants whose terminal buds were killed.

Heat Canker

Heat canker causes girdling of the hypocotyl at or just above the soil line. The affected area varies in length from 0.25 to 1.0 mm, with the affected tissues appearing reddish brown and shriveled. Hypocotyl tissue directly above or below the constriction appears normal. An affected plant may continue to grow for a day or two, elevating the constricted area several millimeters above the soil line (Plate 38). Within a few days the entire plant wilts, shrivels, and dies.

Heat canker is most likely to occur on seedlings that emerge just prior to periods of unusually high temperatures. When air temperatures are near 38°C, surface soil temperatures of 71–77°C are possible. Such temperatures at the soil surface can injure the tender, succulent tissues of recently emerged soybean hypocotyls.

(Prepared by L. E. Sweets)

Lightning

Lightning kills soybeans in circular spots (up to 15 m in diameter) in the field, with a wide border of damaged plants (Fig. 101). The lower parts of stems may be blackened, with many dead leaves still attached (Plate 39). The pith of such plants appears "cooked" or blanched. Lightning damage is sometimes mistaken for a parasitic disease, such as a root rot, but is easily distinguished from such diseases by the sudden death of both soybeans and weeds, the clearly defined margins of the affected area, the lack of any signs of a pathogen, and the fact that the affected area does not expand.

Sunburn

Minor sunburn damage appears as small, interveinal, brick red spots on both leaf surfaces. In severe cases, the discoloration

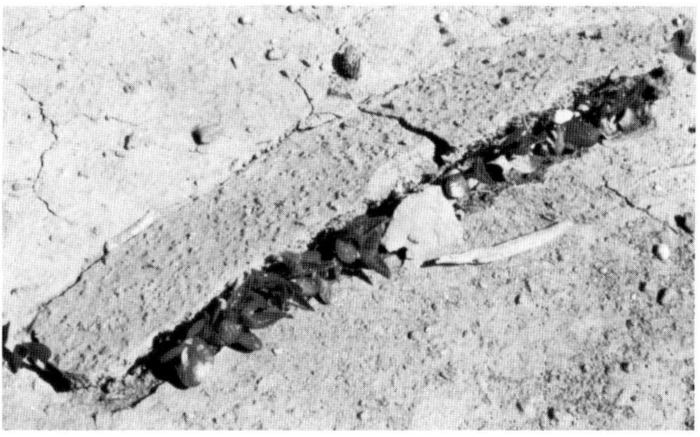

Fig. 100. Crusting of fine-textured soil, retarding the emergence of germinated soybean seeds. (Courtesy W. O. Scott)

spreads over and along the leaf veins (Fig. 102 and Plate 40). The spots later develop brownish centers, which may crack. Sunburn damage on petioles and stems appears as elongated, brick red lesions, which may be confused with bacterial infections. On pods, brown spots appear, spread, and are often colonized later by saprophytic fungi, such as species of *Alternaria* and *Penicillium.*

Water Stress

Soybeans die if they are grown in fields that are flooded or submerged for extended periods. How long plants can survive under such conditions depends on the temperature and whether the water is still or moving. If not submerged, soybeans can generally withstand flooding relatively well. Some cultivars are

Fig. 101. Lightning damage, resulting in a circular pattern of damaged and dead plants. (Courtesy D. W. Chamberlain)

Fig. 102. Sunburn injury on the underside of a soybean leaflet. (Courtesy D. W. Chamberlain)

more flood-tolerant than others. Root injury or death results from an exhaustion of the available oxygen (anoxia) in the soil environment.

Soybeans in low wet areas may die from Phytophthora rot (see Phytophthora Rot) rather than water damage. Herbicide injury may also be potentiated in wet soils.

Water stress can also be caused by drought. Throughout the soybean-growing world, this is probably the single most important yield-limiting factor affecting the crop. The symptoms include stunting, poor vigor, and chlorosis of the interveinal area of leaves, which changes to a general chlorosis with interveinal necrosis in severe situations. Plants in advanced moisture stress may lose many of their leaves. Yield is only slightly affected by drought prior to pod set, if water sufficiency returns later. However, drought during the stages from pod set through pod fill can cause irreversible yield losses. In research trials conducted in Alabama, covering seven years and 35 experiments, rainfall in a 10-day period centered around R_3 was responsible for 40% of the variation in yield observed at harvest. Drought occurring in later growth stages, though less important, can have strong effects on soybean yields.

Damage to roots caused by nematode or fungal diseases can increase sensitivity to drought. Similarly, diseases that block the vascular system or disrupt it by the formation of cankers or galls can also increase drought stress. Many of the symptoms of soybeans with these diseases are due to water stress more than any other aspect of the disease physiology.

The soybean plant must extract its nutrients from the soil solution. As this solution becomes more and more concentrated during drought, shifts in the availability of required nutrients are inevitably produced. Iron and nitrogen are two of the nutrients whose availability is most affected. The symptoms may then also result from the altered nutrient status of the plant.

Selected Reference

Scott, W. O., and Aldrich, S. R. 1983. Modern Soybean Production. S & A Publications, Champaign, IL. 230 pp.

Mineral Deficiencies and Toxicities

Mineral deficiencies are most common in sandy, weathered, highly organic soils exposed to abundant rainfall or heavy irrigation. Soil pH, aeration, moisture, temperature, and microbial activity also affect the availability of nutrients (elements). Changing the soil pH to slightly acid (pH 6.0–6.5) may alter the availability of many elements and reduce the need for direct application of required elements in fertilizers.

Deficiencies and toxicities may be confused with symptoms of infectious disease. Identifying a deficiency or toxicity requires complete information about previous soil treatment. In addition, a soil or plant analysis may be needed for positive diagnosis.

Boron

Soybeans are relatively tolerant of boron deficiency, which occurs in alkaline or strongly acidic soils in areas of low rainfall and in soils high or very low in organic matter. Symptoms of the deficiency include shortened internodes and yellowing or reddening of upper leaves. Flowers fail to develop, particularly in dry weather. Boron deficiency may be confused with leafhopper injury and potassium or calcium deficiency.

Excessive boron, especially near rows and in sandy soils, is extremely toxic to soybeans, especially seedlings. Affected plants are stunted. Toxicity symptoms appear only on leaves, usually those that are just opening (Plate 41). The pubescence at the leaf tip becomes brown. This symptom is followed by browning of the leaf tip and progressive browning of the entire leaf. The leaf continues to develop, but eventually the dead tissue at the tip and along the leaf margins prevents the blade from expanding, the leaf puckers or crinkles, and its edges cup up or down. Older leaves may fall prematurely.

Calcium

Calcium is important for the normal development of the beneficial bacteria *Bradyrhizobium* and for symbiotic nitrogen fixation as well as for the host plant. It affects potassium and magnesium accumulated by plants.

Calcium deficiency is most common in acid, sandy soils low in organic matter. Plants grown in calcium-deficient soils may have nitrogen deficiency symptoms. The emergence of primary leaves is delayed, and they are cup-shaped when emerged. Terminal buds of primary leaves become necrotic, and other parts are chlorotic. Tissues between veins tend to be ridged (Plate 42). Terminal buds wither, and petioles fall. Later, the primary leaves become flaccid and drop prematurely. Plants deficient in calcium are poorly nodulated and more vulnerable to infection by damping-off microorganisms.

Calcium deficiency can be corrected by liming the soil to pH 6.0 or above. Calcium does not neutralize acid soil; a basic anion is required. However, most liming materials that occur in nature contain basic anion compounds.

Copper

Research on copper deficiency in soybeans is limited. The deficiency reduces seed yields, but plants may not show obvious symptoms. It is likely to occur in alkaline peat and muck soils (pH 7–8), inorganic soils high in organic matter, and acid, highly leached, sandy soils. The deficiency can be detected by soil and plant analyses and can be corrected by applying copper sulfate or chelated copper compounds to the soil or by changing the soil pH to slightly acid (pH 6.0–6.5).

Iron

Early leaf symptoms of iron deficiency are a yellowing of interveinal areas on younger leaves (Plate 43). Later, even the veins may become chlorotic; the whole leaf finally turns ivory to white. Brown, necrotic spots may occur near leaf margins. Plant growth is repressed.

Chlorosis caused by iron deficiency is often referred to as lime-induced chlorosis. Chlorotic symptoms may also appear when plants are not using available iron effectively and large amounts of iron accumulate in the leaves. The deficiency is often severe in high-pH soils (usually well above pH 7.0), in which iron salts are converted to the ferric form, which is unavailable to soybean plants.

Well-aerated, alkaline soils low in organic matter, unusually low or high temperatures, and intense sunlight increase the severity of iron deficiency. Excess calcium, copper, manganese, molybdenum, phosphorus, vanadium, or zinc induces iron chlorosis by inhibiting root absorption of this nutrient or interfering with the utilization of it. High concentrations of manganese in poorly aerated soils may inhibit iron absorption and utilization.

Soybean cultivars differ in their tolerance for iron deficiency. At marginal deficiency levels, some cultivars show complete chlorosis, but others remain normal.

Control of iron chlorosis is still being investigated. Foliar application of an aqueous solution of 0.5–1.0% ferric sulfate is suggested as a control measure.

Magnesium

Magnesium, an important plant nutrient, is required for chlorophyll formation, because it is part of the molecule. Magnesium deficiency also inhibits nitrogen fixation by *B. japonicum* (see Beneficial Bacteria).

In early stages of growth, magnesium deficiency is recognized by pale green to yellow interveinal tissues on the leaf blade (Plate 44). Leaves later become deep yellow, except at the base. The lower leaves show symptoms first; later, rusty specks and necrotic blotches may appear between the veins and around the edges of middle and upper leaves. The deficiency gives older plants the general appearance of early maturity. Leaf margins curve down, with a gradual yellowing from the margin inward, followed by bronzing of the entire leaf.

Magnesium deficiency can occur after heavy application of potassium or ammonia to the soil. It is most prevalant in deep, sandy, acidic soils. Application of dolomitic limestone corrects the deficiency. If lime is not needed, sulfate of potash magnesia can be used to supply both magnesium and potassium (and sulfur). Application of magnesium increases the oil content and yield of soybeans and improves growth.

Manganese

Soybeans are sensitive to manganese deficiency, which commonly occurs in plants in well-drained neutral and alkaline soils, such as peat and muck soils. Interveinal areas of leaves become light green to white; the veins remain green. The symptoms resemble the early symptoms of iron deficiency, except the veins of manganese-deficient leaves remain green and stand out prominently (Plate 45). Necrotic, brown spots develop as the deficiency becomes more severe. Leaves may drop prematurely.

The symptoms are pronounced in cool weather; when temperatures rise, affected plants generally outgrow mild deficiency symptoms. Manganese deficiency symptoms frequently appear if the iron concentration is increased.

The deficiency can easily be treated by applying fertilizers that contain manganese to the soil. Aerial application of this nutrient to young deficient plants can increase soybean yields. The element is best applied as a spray of manganese sulfate when plants are 15–25 cm tall; one application usually is sufficient. Another good way to correct the deficiency, especially recommended for soils high in organic matter, is to mix manganese salts with acid-forming fertilizers, such as superphosphate, and apply the mixture in bands beneath the rows. The availability of manganese is effectively increased by combining it with phosphorus and potash; with nitrogen, phosphorus, and potash; or with nitrogen and potash. Manganese must be applied yearly, because alkaline soils bind it in an unavailable form.

Manganese toxicity appears in plants growing in soils below pH 5.0 containing abundant iron-manganese pebbles and is associated with high soil moisture and management practices that increase soil acidity. Plants are stunted, and leaves are crinkled and cup down (Plate 46). The youngest leaves are affected first. Yellowing between leaf veins may or may not be observed. The toxicity can be corrected by liming the soil to pH 6 or slightly above.

Molybdenum

Molybdenum is required for nitrogen metabolism in leaf tissues and for nitrogen fixation by *B. japonicum* (see Beneficial Bacteria). Molybdenum deficiency occurs in highly acidic soils that are strongly weathered and leached and in soils in which the element is in an unusable form. The deficiency greatly reduces yields. Plant growth, the number of pods, the number of seeds per pod, seed size, nodulation, and the total nitrogen and protein contents of the seeds are reduced.

Symptoms of this deficiency resemble those of nitrogen deficiency and are probably caused indirectly by reduced nitrogen utilization rather than directly by lack of molybdenum. Leaves are pale green or yellow, necrotic, and twisted. The necrosis is largely confined to the midribs, interveinal areas, and margins.

The deficiency can also arise when seeds from plants grown in molybdenum-deficient soil are planted. Seeds with a low molybdenum content may produce plants that show deficiency symptoms when grown in soils with a marginal supply of this nutrient.

Molybdenum deficiency can be corrected by treating seeds with sodium molybdate or ammonium molybdate. Foliage sprays with these salts at the prebloom stage are also effective. In many soils, the deficiency can be remedied by liming the soil to pH 6.0–6.3.

Nitrogen

Nitrogen-deficient plants become pale green. Later, the leaves turn distinctly and uniformly yellow (Plate 47). Symptoms appear first on the basal leaves and quickly spread to upper parts. The plants eventually defoliate and are often stunted and spindly. The deficiency can be diagnosed by analyzing leaves for nitrogen, uprooting the plant to check for nodule formation by *Bradyrhizobium*, and analyzing the soil to determine its pH and calcium content.

Soybeans normally do not need nitrogen fertilization, because the requirement for this nutrient is met by the nodulating bacterium *B. japonicum* and by available soil nitrogen. The deficiency occurs when soil nitrogen is low, soil moisture is limiting, and molybdenum is deficient. Where soybeans are grown for the first time, seed inoculation with *B. japonicum* is required. Occasionally, nitrogen deficiency symptoms are caused by the soybean cyst nematode or other nematodes, root rot, or molybdenum deficiency, all of which interfere with nodule development.

Phosphorus

Soybeans require relatively large amounts of phosphorus, especially at pod set. It is required for normal nitrogen fixation by *B. japonicum*.

Symptoms of phosphorus deficiency are not well defined. The chief symptom is retarded growth; affected plants are spindly and have small leaflets. Leaves turn dark green or bluish green. The leaf blade may curl up and appear pointed. Blooming and maturity are delayed. The dark green of the leaves gives the impression that the plants are quite healthy.

Leaf analysis for phosphorus content is conducted for diagnosis. Soil analyses for available phosphorus and soil pH are also good indicators of the availability of the nutrient. In acidic soils, it is in a form not available to soybean roots. Phosphorus uptake is reduced in cool, wet soils.

The deficiency can be corrected by the application of fertilizers containing phosphorus. This is profitable when the availability of the nutrient in the soil, as determined by a soil test, is low and other nutrients are in adequate supply. The fertilizer should be well mixed into the soil, because this element moves slowly through most soils. Application of phosphorus increases the number and weight of nodules and thus increases the nitrogen-fixing capacity of nodulating bacteria.

Mycorrhizal fungi associated with soybean roots may help overcome phosphorus deficiency.

An elevated phosphorus content in the soil may induce zinc and iron deficiencies if those elements are already on the border line of short supply.

Potassium

Soybeans require large amounts of potassium. It is important for all aspects of plant growth and influences the plant's nutritional balance. It is also involved in the uptake of calcium and magnesium. The maximum need for potassium uptake occurs during periods of rapid vegetative growth. Plants deficient in this nutrient tend to have weak stems and are more susceptible to some plant disease.

Potassium deficiency symptoms are well defined. In early stages of growth, an irregular yellow mottling appears around leaflet margins. The yellow areas coalesce to form a more or less continuous, irregular yellow border (Plate 48). This symptom is often followed by necrosis of chlorotic areas and downward cupping of leaf margins (Plate 49). The dead tissues then drop away, so that the leaves acquire a ragged appearance. Chlorosis and necrosis may spread inward to include half or more of the leaflet. The basal portions remain green. Symptoms appear first on the older leaves. Severe deficiency tends to produce wrinkled and misshapen seeds.

Injury from leafhoppers feeding on cultivars with glabrous leaves often mimics potassium deficiency.

The deficiency can be corrected by applying fertilizers that contain potassium, as determined by a soil test. This increases yield, by increasing the number of pods and the number, size, and weight of seeds. It also increases nodulation by *Bradyrhizobium*.

Sulfur

The main symptom of sulfur deficiency is small, yellow-green leaves. Stems are thin, hard, and elongated. Leaf symptoms resemble those produced by deficiencies of other elements, such as nitrogen and phosphorus, but stem elongation is characteristic of sulfur deficiency.

The deficiency results in the accumulation of carbohydrates, nitrate, and amino acids that do not contain sulfur. The synthesis of sulfur-containing amino acids is interrupted; thus, protein synthesis is reduced, and proteolysis is increased. Cell walls thicken from the accumulation of starch and hemicellulose.

The availability of sulfur depends on the rate at which it is released from organic matter, which in turn is influenced by the kind of plant residues, soil moisture, and soil pH. Sulfur deficiency can be corrected by applying fertilizers that contain sulfur to the soil.

Zinc

Soybean yields are considerably reduced in zinc-deficient soils. Deficient plants have stunted stems and leaves with chlorotic interveinal areas (Plate 50). Later, entire leaves turn yellow or light green. Lower leaves may turn brown or gray and drop early. Few flowers are formed, and the pods that set are abnormal and slow in maturing. Yield reduction is primarily the result of the reduced number of seeds formed.

Zinc deficiency is most likely to occur in strongly weathered, coarse-textured soils that are alkaline (with free lime); in eroded soils; or in soils excessively fertilized with phosphorus and low in organic matter. Adding lime and phosphorus may reduce zinc availability and thus cause a deficiency of this nutrient.

The deficiency can be corrected by applying a zinc salt (most commonly zinc sulfate or zinc oxide) to the foliage or soil. It may be broadcast and worked in or placed in a band at least 5 cm to the side and 5 cm below the seed. A single broadcast application is generally enough for 2 to 4 years. Seed treatment with zinc salts is also helpful.

Soybeans are also sensitive to zinc toxicity, particularly in fields near industrial plants that emit large amounts of zinc or after heavy applications of zinc salts, particularly in acidic soils. The toxicity can be corrected by liming soil to pH 6.2–6.5.

Soybean cultivars vary greatly in their tolerance for zinc deficiency and toxicity.

Selected References

Raper, C. C., Jr., and Kramer, P. J. 1987. Stress physiology. Pages 590-641 in: Soybeans: Improvement, Production, and Uses, 2nd ed. J. R. Wilcox, ed. American Society of Agronomy, Madison, WI.

Scott, W. O., and Aldrich, S. R. 1983. Modern Soybean Production. S & A Publications, Champaign, IL. 230 pp.

Sprague, H. E., ed. 1964. Hunger Signs in Crops: A Symposium, 3rd ed. D. McKay, New York. 461 pp.

Herbicide Damage

It is often difficult to apply enough herbicide for effective weed control without risking crop injury. Application of a postemergence herbicide may injure plants stressed by cool, wet weather or hot, dry conditions. Soybeans may also be damaged by herbicide carryover, excessive rates of application, application of the wrong chemical, overlaps, uneven application, improperly operated equipment, or incorrect incorporation of a chemical herbicide.

In most cases, soybean plants outgrow herbicide damage with little or no loss of yield. Sometimes, however, stands or plant vigor may be reduced. Soybean cultivars vary in their susceptibility to particular herbicides. Poor-quality seeds tend to produce weak seedlings, which are more vulnerable to herbicide injury.

Chemical injury usually occurs in characteristic patterns in a field, often associated with soil types or the movement of application or incorporation equipment. If herbicide damage is suspected, root systems should be examined as well as above-ground parts. The roots often hold the key to diagnosis. If they are deformed, chemical injury is likely. Soybean growth abnormalities associated with various herbicides are described in Table 13 and illustrated in Plates 51–60.

Residues from herbicides applied to the previous crop in a field may injure soybean seedlings. In a field with triazine residues, for example, seedlings emerge and appear normal for a week or longer; then, as the roots grow and take up more residual herbicide, the leaves become chlorotic and brown. Affected plants may outgrow the damage or die, depending on the amount of residual herbicide present. Injury from atrazine residues is typical of this type of damage (Table 13). Herbicide residues are most likely to appear where the soil is low in organic matter or high in pH.

Soybean plants may be damaged by drifting spray or vapor of some herbicides, such as 2,4-D and dicamba, used in fencerows and along roadsides for weed control. Drift damage can usually be distinguished from infectious disease by the pattern of injury in the affected field. All plants in the area exhibit identical symptoms, and the damage is usually most severe around the margins of the field closest to where the spraying occurred. If weeds or other susceptible crops in the area show similar symptoms, herbicide drift is probably the cause.

Herbicide damage may be avoided by measuring and applying herbicides at the correct rates, at the proper time and growing stage, and in a uniform manner.

(Prepared by M. D. McGlamery and W. S. Curran)

Insecticide Damage

When properly applied as indicated on the labels on their containers, insecticides rarely cause phytotoxic injury to soybeans. However, certain combinations of soil-applied insecticides and herbicides can injure seedlings, and certain foliage-applied insecticides can cause phytotoxicity under some

Table 13. Symptoms and Control of Herbicide Injury to Soybeans

Herbicide	Symptoms	Control	Remarks
Alachlor Metolachlor	Seedling leaves are slightly crinkled. Leaflet tips appear indented; leaflets have a shortened midrib and are heart-shaped.	Use the correct rate and apply carefully. Plant high-quality seeds.	The injury is usually temporary, with no significant reduction in vigor or yield.
Bentazon	Bronzing or splash burn.	Use oil only when necessary to increase the effectiveness of weed control.	The injury increases with previous stress or injury and with the use of oils.
Chloramben	Seedlings are stunted from stubbing, proliferation, and malformation of roots (Plate 51).	Proper rates and methods of application are important. Avoid deep planting, and plant high-quality seeds.	Plants usually recover without significant loss of vigor or yield. The injury is often associated with moisture that moves the herbicide to the seed zone soon after planting.
Dicamba	Young leaves are cupped and crinkled, with shortened veins, and internodes of stems are dwarfed (Plates 52 and 53); the symptoms are similar to those of 2,4-D injury. Dicamba injury resembles soybean mosaic, except for the cupping of leaves.	Avoid drift. Apply to other crops with ground application and drift inhibition techniques.	The injury is usually the result of spray drift and hence is found along roads, fencerows, and rights-of way, next to fields of maize or small grains, and next to pastures.
Dinitroanilines Ethalfluralin Trifluralin Oryzalin Pendimethalin	Plants are stunted and slow to emerge, hypocotyls are enlarged, and secondary roots are limited, thick, and stubby (Plate 54). Leaves may be small and crinkled. The injury may be significant in fields where cool, wet conditions prevail, overlapping has occurred, or the rate was not adjusted for the soil texture and the organic matter content of the soil.	Adjust the rate for the soil type. Apply accurately and uniformly and incorporate well enough for uniform distribution. Do not exceed the recommended rate.	Thickened hypocotyls may also be produced when crusting or compacted soil interferes with seedling emergence, but in these cases, root systems are normal. Dinitroaniline injury may increase susceptibility to root rot. Soybeans under stress from disease, deep planting, or cold, wet weather are often more susceptible to injury from these compounds.
Diphenyl ethers Acifluorfen Fomesafen Lactofen	Contact burn accentuated by surfactants or crop oil concentrates.	Apply accurately. Use additives only if needed.	These compounds are contact herbicides. The injury increases with temperature and stress.
Imazaquin Imazethapyr	Plants are stunted and have shortened internodes, golden terminal leaves, and shoots from lower buds.	Apply and incorporate accurately and uniformly.	The injury occurs most frequently on light-textured soils with low organic matter.
Chlorimuron-ethyl DPX M-6316	Plants are stunted and have shortened internodes, golden terminal leaves, and shoots from lower buds.	Apply and incorporate accurately and uniformly.	The injury occurs most frequently on light-textured soils with low organic matter.
Clomazone	Whitening or bleaching of leaves and plants.	Apply and incorporate accurately and uniformly.	Soybeans are very tolerant of clomazone.

(continued on next page)

Herbicide	Symptoms	Control	Remarks
Linuron	Plants are chlorotic; seedlings may die back. Leaf margins and tips are "scorched" and yellow to brown (Plate 55). In severe cases, entire leaves turn yellow to brown. The stand may be spotty and uneven.	Adjust the rate for the soil type, and apply accurately. Avoid overlap. Use other herbicides on fields with variable soils. Do not apply on sandy soils.	The injury may be mistaken for seedling blight. It occurs most commonly on sandy soils high in organic matter. The margin of selectivity is rather narrow.
Metribuzin	The injury is difficult to distinguish from linuron injury or triazine carry-over.	Adjust the rate for soil conditions. Apply accurately. Reduce the rate in herbicide combinations. Do not apply on sandy soils.	Metribuzin may have an additive effect with atrazine or simazine carryover. Even at the recommended rate, metribuzin may injure soybeans if used with a soil-applied organophosphate insecticide (Plate 56). Cultivars may differ in susceptibility.
Triazines Atrazine Simazine	Carryover causes progressive yellow mottling of leaves, followed by dead areas in the leaves of young plants (Plate 57). The symptoms are similar to those of linuron and metribuzin injuries. In severe cases, entire plants turn yellow to brown, dry out, and die.	Do not exceed the recommended rate, overlap, or apply excessive amounts. Metribuzin may accentuate the injury.	Carryover injury occurs where soybeans follow a crop to which atrazine or simazine was applied the previous year. It is most prevalent after a cool, dry year and on high-pH soils. Plants usually recover from initial injury.
2,4-D	Young leaves are thickened, crinkled, and paddle- or strap-shaped, with parallel veins (Plate 58). The injury is often observed in a diminishing pattern, according to spraying operations and drift.	Avoid spraying 2,4-D near soybean fields, especially on windy days. The amine form is safer than the ester. Increase the average drop size and reduce mist spray. Aerial application requires extreme caution.	The injury usually occurs along roads and fencerows or next to other fields that were sprayed. It is sometimes confused with soybean mosaic or other viral diseases.
2,4-DB	Plants may have curved or cracked stems, crinkled leaves, and callus tissue on the lower stem, and lodging may occur. Adventitious roots may form along stems (Plate 59).	Use only for emergency weed control. Apply the recommended rate accurately. Directed sprays are safer than topical applications. Do not spray more than the lower third of soybean plants.	This compound is used in established soybeans to control cockleburs and some other weeds. Injury is most likely under hot, dry conditions.
Vernolate Butylate	Emergence is delayed, plants are stunted, and leaves are slow to expand. Hypocotyls may be enlarged (Plate 60).	Do not exceed the suggested rate. Apply and incorporate accurately and uniformly.	Plants usually recover from initial injury without significant yield reduction unless secondary stress occurs.

environmental conditions. As with herbicide injury, soybean plants usually outgrow insecticide damage with little or no subsequent loss of yield.

Soybeans may be injured if the herbicide metribuzin is used with a soil-applied organophosphate insecticide. Organophosphates labeled for use on soybeans to prevent insects and nematodes from attacking planted seed and underground portions of the plants include chlorpyrifos, diazinon, ethoprop, and phorate. These compounds and metribuzin used in combination cause symptoms of metribuzin injury (see Table 13): plants are chlorotic, seedlings may die back, leaf margins and tips are scorched and yellow to brown, in severe cases the entire leaf turns yellow to brown, and the stand may be spotty and uneven (Plate 56). The insecticide apparently makes metribuzin more active, even when the latter is applied at the rate recommended on the label. This injury is most likely to occur in soybeans grown from poor-quality seed or in seedlings under environmental stress. Prevention of the injury is simple: soil-applied organophosphates and metribuzin should not be applied in the same field.

The labels of soil-applied organophosphate insecticides also state that they should not be applied in the seed furrow or allowed to contact the seed. Insecticides applied incorrectly in cool, wet weather, during which soybeans grow slowly, can result in seedling injury. The emergence of seedlings may be delayed, and injured plants may be stunted.

Some formulations of certain insecticides may cause spot burning if they are applied to tender soybean foliage that is wet or applied during periods of high humidity.

The labels of many insecticides also prohibit foliar application in combination with certain herbicides. Injuries resulting from these combinations resemble those caused by the herbicides alone.

(Prepared by K. L. Steffey and D. E. Kuhlman)

Air Pollutants

Many airborne chemicals are known to injure soybean leaves. The symptoms are diverse and often mimic those caused by other factors. Thus, the precise cause of foliar symptoms is usually difficult to determine. Injury caused by most air pollutants is infrequent and limited to areas near the source of emission. For example, pesticides can be considered air pollutants in cases of injury caused by drift of compounds such as carbaryl (Plate 61) and dicofol (Plate 62) from nearby fields. Injury from chlorine (Plate 63), hydrogen chloride, and ammonia (Plate 64) can occur after accidental spills. Recurring injury can affect soybeans near industrial sources of fluorides and sulfur dioxide, but recent control strategies involving scrubbers and tall stacks have greatly decreased this problem. Ozone is now considered the most important phytotoxic air pollutant, because it regularly occurs regionally at levels injurious to many crops, including soybeans.

Ammonia

Leaves that are almost fully expanded are most sensitive to ammonia. The injury ranges from local, reddish brown lesions to broad, cream to brown bands between veins (Plate 64). The pollutant source must be known in order to distinguish ammonia damage from that caused by other pollutants.

Fluorides

The most common symptom of fluoride injury to soybeans, as with most other crops, is marginal chlorosis and necrosis of older leaves. Knowledge of a nearby fluoride source and foliar analysis for fluoride content are required for diagnosis.

Ozone

Most of the ozone in rural areas is formed by complex photochemical reactions involving nitrogen oxides and hydrocarbons transported from major urban areas. The daily exposure level is determined by the concentrations of hydrocarbons and nitrogen oxides in the air and the amount of sunshine. Seasonal exposures consist of days when ozone levels are relatively low or near average, punctuated by days when levels are high. Concentrations in rural areas are often higher than in urban areas.

Soybean leaves that have recently reached full expansion are more sensitive to ozone than younger leaves. Growing conditions that favor gas exchange between the leaves and the air also favor ozone uptake by the leaves. Symptoms of the injury may include fairly discrete upper-surface flecking, producing white to tan necrotic areas less than 1 mm in diameter, or upper-surface stippling, producing dark areas less than 1 to 2 mm in diameter (Plate 65). More commonly, however, the only foliar symptoms are an increased rate of chlorosis, necrosis, and abscission, which are virtually indistinguishable from symptoms of normal senescence. Field research estimates of soybean yield losses caused by ozone range from 4 to 23%, depending on the cultivar, season, and location.

Sulfur Dioxide

Symptoms of sulfur dioxide injury include chlorosis and irregular cream-colored to light brown necrotic areas between leaf veins. As with ozone, repeated exposure to sulfur dioxide in low concentrations may accelerate the senescence of leaves. An example of unusually severe symptoms from an experimental exposure is shown in Plate 66. Yields may be reduced if soybeans are exposed to large amounts of sulfur dioxide after flowering. The source of the pollutant, burning high-sulfur fuel or smelting high-sulfur ore, is usually close by. Sulfur dioxide may increase soybean yields in areas where soils are low in sulfur, but research has not verified this possibility.

Dicofol, a miticide, usually damages fully expanded soybean leaves. Dead areas resembling tissue injured by sulfur dioxide appear between veins and at leaf margins (Plate 62). Knowledge of what chemicals were applied is essential to a proper diagnosis.

Selected References

Heagle, A. S., Heck, W. W., Rawlings, J. O., and Philbeck, R. B. 1983. Effects of chronic doses of ozone and sulfur dioxide on injury and yield of soybeans in open-top field chambers. Crop Sci. 23:1184-1191.

Heck, W. W., Heagle, A. S., and Shriner, D. S. 1986. Effects on vegetation: Native, crops, forests. Pages 248-333 in: Air Pollution, 3rd ed., Vol. VI: Supplement to Air Pollutants, Their Transformation, Transport and Effects. A. C. Stern, ed. Academic Press, New York.

Jacobson, J. S., and Hill, A. C., eds. 1970. Recognition of Air Pollution Injury to Vegetation: A Pictorial Atlas. Air Pollution Control Association, Pittsburgh, PA. 110 p.

Laurence, J. A., and Weinstein, L. H. 1981. Effects of air pollutants on plant productivity. Annu. Rev. Phytopathol. 19:257-271.

Noggle, J. C., Meagher, J. F., and Jones, H. C. 1986. Sulfur in the atmosphere and its effects on plant growth. Pages 251-278 in: Sulfur in Agriculture. M. A. Tabatabai, ed. Agron. Ser. No. 27. American Society of Agronomy, Crop Science Society of America, and Soil Science Society of America, Madison, WI.

(Prepared by A. S. Heagle)

Part IV.
Soybean Disease Management Strategies

Soybean diseases reduce yields by 10–30% in most production areas. These losses are generally caused by several diseases rather than any single one. Disease losses can be reduced by the implementation of a comprehensive disease management program that addresses the principles of disease control: exclusion, eradication, protection, and the use of disease-resistant cultivars. The choice of a particular strategy should be dependent on an assessment of disease loss potential, the cost and benefits of the strategy, and the long-term environmental impact.

Exclusion

There are two major approaches to disease control by exclusion: legislation and the use of disease-free seed. Exclusion is useful where a pathogen does not exist, cannot survive between crops, or is not present in an area and is not likely to be introduced as airborne inoculum.

Legislative embargoes, quarantines, or inspections are widely used to prevent the introduction of exotic plant pathogens, weeds, or insects into new areas. These legislative actions limit the spread of pathogens that do not occur worldwide but pose a significant risk to soybean production. Knowledge of epidemiology, the life cycle and distribution of the pathogen, and the risk it poses for production in a particular environment is needed if legislative embargoes, quarantines, or inspections are to be effective.

The use of disease-free seed is an effective means of preventing the entrance of seedborne pathogens. Disease-free seed can be produced in an area free of a particular disease or isolated from it; in an area with low rainfall, to exclude pathogens spread by splashing water; or in an area free of vectors of particular virus diseases. It can also be obtained from seed production fields or seed lots inspected to ensure the absence of a pathogen or a lack of pathogen activity. Seeds can also be treated with fungicides or bactericides or can be produced from plants treated with fungicide or insecticide sprays. Regulatory measures include outright embargo of seeds from certain countries and field inspections or laboratory tests by an official agency to certify freedom from certain pathogens either in the production field or in seeds being stripped.

Important soybean diseases whose spread is limited by exclusion and regulated by one or more countries include rust, the soybean cyst nematode, downy mildew, bacterial blight, bacterial pustule, bacterial wilt, anthracnose, purple seed stain, frogeye leaf spot, peanut stripe virus, tobacco ringspot virus, and viruses known to be seedborne.

Selected Reference

Anonymous. n.d. Export Inspection Manual for Quarantinable Diseases of Soybeans (*Glycine max*). Plant Protection Quarantine, Animal and Plant Health Inspection Service, U.S. Department of Agriculture, Hyattsville, MD. 37 pp.

Eradication

Eradication involves methods that either eradicate or reduce inoculum or reduce the effectiveness of the inoculum. In soybean production, crop rotation and plowing or tillage to destroy crop residues where pathogens overseason are the primary methods.

Rotation

Crop rotation is a very powerful tool for the management of soybean diseases, particularly if it involves members of the grass family, such as maize (*Zea mays* L.), sorghum (*Sorghum bicolor* (L.) Moench), wheat (*Triticum aestivum* L.), barley (*Hordeum vulgare* L.), rye (*Secale cereale* L.), or oats (*Avena sativa* L.), since they are hosts of only a very few soybean pathogens. Rotation with resistant cultivars or crops that reduce inoculum buildup can also be effective. Examples include rotation with resistant cultivars to control the soybean cyst nematode and brown stem rot and rotation with grain sorghum or cotton (*Gossypium* sp.) to reduce the severity of charcoal rot. When these crops are planted for one or more seasons in rotation with soybeans, many soybean pathogens die out, or their inoculum levels are sharply reduced. Diseases that can be reduced by rotation are listed in Table 14.

The economics of rotation over several years should be considered, not just the value generated in a given year. There are numerous examples of soybean yield increases after rotation with nonlegumes, even where diseases do not appear to be limiting. These increases may be due to suppression of disease, improved weed control, soil tilth, or other factors.

Selected References

Ferris, V. R., and Barnard, R. L. 1971. Crop rotation effects on populations of ectoparasitic nematodes. J. Nematol. 3:119-122.

Melton, T. A., Edwards, D. I., and Noel, G. R. 1988. The soybean cyst nematode problem. Univ. Ill., Dep. Plant Pathol., Rep. Plant Dis. No. 501. 13 pp.

Moore, W. F., ed. 1981. Soybean Cyst Nematode. Soybean Industry Resource Committee, Extension Service, U.S. Department of Agriculture, Washington, DC. 23 pp.

Riggs, R. D. 1985. Strategies for race stabilization in soybean cyst nematode. Pages 528-538 in: Proc. World Soybean Res. Conf. III. R. Shibles, ed. Westview Press, Boulder, CO.

Tachibana, H. 1988. Use and management of resistance for control of brown stem rot of soybeans. Pages 102-105 in: Soybean Diseases of the North Central Region. T. D. Wyllie and D. H. Scott, eds. American Phytopathological Society, St. Paul, MN.

Welch, L. F. 1985. Rotational benefits to soybeans and following crops. Pages 1054-1060 in: Proc. World Soybean Res. Conf. III. R. Shibles, ed. Westview Press, Boulder, CO.

Wyllie, T. D. 1988. Charcoal rot of soybeans—Current status. Pages 106-113 in: Soybean Diseases of the North Central Region. T. D. Wyllie and D. H. Scott, eds. American Phytopathological Society, St. Paul, MN.

Tillage

Destruction of crop residues by tillage or other practices can be an effective means of reducing overwintering inoculum of most foliage- and stem-infecting bacteria and fungi attacking soybeans. The majority of these pathogens survive between soybean crops on or in infested residues, and when these residues are forced to decay following tillage, the pathogens cannot compete with soil microorganisms and die out. The effect of the destruction of crop residues by tillage on various soybean diseases is shown in Table 14.

Tillage can also influence compaction and thereby drainage of the soil. Systems that reduce compaction can reduce the incidence and severity of Phytophthora root rot.

Selected References

Bowman, J. E., Hartman, G. L., McClary, R. D., Sinclair, J. B., Hummel, J. W., and Wax, L. M. 1986. Effects of weed control and

Table 14. Relative Effectiveness of Selected Strategies for Control of Principal Soybean Diseases[a]

Disease	Cultural Practice[c]	Resistant Cultivars[d]	Crop Rotation	Tillage	High-Quality Seeds	Pesticides[e]
Bacterial diseases		1	2	2	2	
Bacterial blight	CT, F	1	2	2	2	
Bacterial crinkle leaf spot	CT					
Bacterial pustule	CT, F	1	2	2	2	
Bacterial tan spot	CT	1				
Wildfire	CT		2	2	2	
Mycoplasmalike diseases						I
Fungal diseases						
Fungal diseases of foliage, upper stems, pods, and seeds						
Alternaria leaf spot			2	2	2	2
Anthracnose		Pi	2	2	2	1 ST, F
Brown spot			2	2		1 F
Cercospora leaf spot and blight		1	2	2	3	1 F
Choanephora leaf blight	A		2	2		2 F
Downy mildew		2			3	
Frogeye leaf spot		1	2	2	3	1 F
Phyllosticta leaf spot			2	2	3	
Powdery mildew		1				2 F
Red leaf blotch	A	1	2	2		2 F
Rhizoctonia aerial, foliage, and web blight	A	1	3	3		1 F
Rust		Pi	3	3		2 F
Scab		Pi	3	3		
Target spot		1				
Yeast spot					3	2 I
Fungal diseases of roots and lower stems						
Brown stem rot		1	1			
Charcoal rot	F, PD, W		2	3		
Fusarium blight or wilt, root rot, and pod and collar rot	F, CT	2[f]			3	3 N
Neocosmospora stem rot						2 N
Phytophthora rot	F, W	1		3		2 ST, SF
Pod and stem blight		Pi	2	2	2	2 F, ST
Pythium rot	PD, W	3			2	1 ST, SF
Red crown rot	A					
Rhizoctonia root rot, stem decay, and damping-off	CT, PD, F				2	3 SF
Sclerotinia stem rot		1	2	3	3	
Sclerotium blight (southern blight)	CT		2			
Stem canker	F, W	1	2	3	3	2 F, 3 ST
Thielaviopsis root rot		2				
Virus diseases						
Bean pod mottle	WC					3 I
Brazilian bud blight	WC					3 I
Bud blight	WC				3	3 I
Cowpea chlorotic mottle		1				3 I
Peanut mottle		1	3			
Soybean mosaic		1			3	
Yellow mosaic		1				
Nematode diseases						
Reniform nematode		1				2 N
Root-knot nematodes		1	2			2 N
Lesion nematodes		2	3			2 N
Soybean cyst nematode	W, F	1	2			2 N

[a] For diseases omitted from the table, no control strategies are available from descriptions in the literature.
[b] Numerical ratings for relative effectiveness of practices: 1 = highly effective; 2 = moderately effective; 3 = slightly effective.
[c] A = increased ventilation (e.g., row spacing, plant density); CT = cultivation techniques; F = optimal fertility; PD = plant density; W = water management; WC = weed control.
[d] Pi = resistance identified but not in commercial varieties.
[e] F = fungicide; I = insecticide; N = nematicide; SF = soil fungicide; ST = seed treatment.
[f] Cultivars resistant to *Fusarium* and to soybean cyst and root-knot nematodes.

row spacing in conventional tillage, reduced tillage, and nontillage on soybean seed quality. Plant Dis. 70:673-676.

Cook, R. J., Boosalis, M. G., and Doupnik, B. 1978. Influence of crop residues on plant diseases. Pages 147-163 in: Crop Residue Management Systems. W. R. Oschwald, ed. Spec. Publ. 31. American Society of Agronomy, Madison, WI.

Minton, N. A., and Parker, M. B. 1987. Root-knot nematode management and yield of soybean as affected by winter cover crops, tillage systems, nematicides. J. Nematol. 19:38-43.

Rothrock, C. S., Hobbs, T. W., and Phillips, D. V. 1985. Effects of tillage and cropping systems on incidence and severity of southern stem canker of soybean. Phytopathology 75:1156-1159.

Schmitthenner, A. F., and Van Doren, D. M., Jr. 1985. Integrated control of root rot of soybean caused by *Phytophthora megasperma* f. sp. *glycinea*. Pages 263-266 in: Ecology and Management of Soilborne Plant Pathogens. C. A. Parker, A. D. Rovira, K. J. Moore, P. T. W. Wong, and J. R. Kollmorgen, eds. American Phytopathological Society, St. Paul, MN.

Sumner, D. R., Doupnik, B., Jr., and Boosalis, M. G. 1981. Effect of reduced tillage and multiple cropping on plant diseases. Annu. Rev. Phytopathol. 19:167-187.

Protection

Protection techniques include the use of pesticides (such as fungicides for treating seed, soil, or foliage; nematicides; and insecticides for vector control) and cultural practices for disease control. Cultural controls include maintaining optimal availability of plant nutrients and optimal soil pH, avoidance of water excess or deficit, planting high-quality seed under conditions that favor rapid germination and emergence, weed control, effective inoculation with *Bradyrhizobium japonicum*, planting nonhost buffer crops to prevent virus spread from field borders, planting optimal populations, harvesting at maturity, and maintaining storage temperatures and equilibrium relative humidities that slow or preclude the growth of storage fungi. Diseases controlled by pesticides and by cultural practices are listed in Table 14.

Seed Treatment Fungicides

Fungicides can be used as seed treatments, soil treatments, or foliar sprays. Seed treatments are generally not needed where high-quality seed is planted at an appropriate seeding rate under conditions that favor rapid germination and emergence. Also, they are not useful where seed quality is low because of mechanical damage or physiological factors. However, they are useful where seed quality is low because of fungal infection, where seeds must be planted in seed beds where germination will be delayed, or where low seeding rates must be used. Seed treatments can effectively reduce losses from seedborne *Phomopsis* sp., *Pythium* sp., and *Rhizoctonia solani* and losses from seedling damping-off induced by *Phytophthora megasperma* f. sp. *glycinea*.

Seed treatments are available as dusts, flowable suspensions, liquids, or wettable powders and can be applied as liquid slurries or dusts. They can be applied in the planter box or prior to planting by machine. When machines are used for treating seed, it is important to avoid mechanical damage. Such damage can be particularly severe in the handling of seeds with moisture contents of less than 10-11%.

Bradyrhizobium inoculants and seed treatment fungicides. The effectiveness of inoculation with *B. japonicum* can be reduced if the inoculant is applied to fungicide-treated seeds in advance of planting. Considerable research has identified fungicides that adversely affect bacterial inoculants and conditions under which the inoculants can be used effectively with fungicides. This research indicates that inoculants applied in a peat moss base with water immediately before planting are compatible with most commonly used fungicides. However, the viability of *Bradyrhizobium* inoculants is reduced, to a point at which nodulation is reduced, if they are applied more than 10-12 hr in advance to seeds treated with several common fungicides.

Fungicides that consistently reduce nodulation include copper-based compounds, oxycarboxin, and captan. Benzimidazoles (benomyl, carbendazim, and thiabendazole), thiram, and dithiocarbamates (maneb, zineb, and mancozeb) have little or no effect, even after long periods of contact. Reports on the effects of pentachloronitrobenzene (PCNB, or quintozene), carboxin, chloranil, dichlone, thiram, and 2-(thio-cyanomethylthio)benzothiazole (TCMTB) are contradictory.

Selected References

Agarwal, V. K., and Sinclair, J. B. 1981. Seed dressings and *Rhizobium inoculum*. Pages 127-131 in: Soybean Seed Quality and Stand Establishment. J. B. Sinclair and J. A. Jackobs, eds. INTSOY Ser. 22. College of Agriculture, University of Illinois at Urbana-Champaign.

Backman, P. A., and Carter, L. 1979. A method for evaluating effects of seed treatment fungicides to *Rhizobium*. Proc. Lat. Am. Congr. Phytopathol., 1st, pp. C1-11

Ferriss, R. S., Stuckey, R. E., Gleason, M. L., and Siegel, M. R. 1987. Effects of seed quality, seed treatment, soil source, and initial soil moisture on soybean seedling performance. Phytopathology 77:140-148.

Henning, A. A. 1988. Seed treatments and the fungal pathogens they are designed to control. Pages 14-21 in: Soybean Diseases of the North Central Region. T. D. Wyllie and D. H. Scott, eds. American Phytopathological Society, St. Paul, MN.

Foliage-Applied Fungicides

Foliage-applied fungicides are commonly used to increase yields and improve seed quality through the control of anthracnose, Cercospora leaf blight, purple seed stain, frogeye leaf spot, Septoria brown spot, and pod and stem blight. Where warm, wet conditions (with temperatures above 25°C) prevail during growth stages R_1 through R_5, these diseases can reduce yields by 10-15% or more and can cause serious seed quality problems. In general, fungicides applied at growth stages R_2 through R_5 provide maximal yield increases and improve seed quality by reducing fungal infection. The yield increases are primarily due to larger average seed size. With increasingly more restrictive grading standards, reduced fungal infections may increase the value of grain lots, and fungicide treatment may thereby become more economical. Fungicides applied at stage R_6 or later produce little if any yield increase but significantly reduce seed infection with anthracnose, purple seed stain, and pod and stem blight, thereby increasing seed vigor and germination.

Researchers in several regions of the United States have developed predictive systems to aid both grain and seed producers in making decisions about fungicide application. Some of these are point systems based on field scouting, forecasts or historical weather data (or both), cropping history, yield potential, planting date, and maturity group (Tables 15 and 16).

Another system is a simple model developed at Auburn University for the management of mid-season diseases (frogeye leaf spot, anthracnose, and brown spot) of soybeans in the southern United States. It is based on a required yield expectation and meteorological records of the number of days with measurable rain, extended periods of fog, or heavy dews. After growth stage R_1, if such events are recorded on 2 or more days in a 4-day period and are predicted for 2 of the next 5 days, an application of a systemic fungicide (benomyl) is recommended. If the fungicide is applied, weather records need not be kept for the next 8 days. After that period, a second application may be made if such events occur again on 2 days during a 4-day period and are predicted for 2 of the next 5 days. In this manner, up to three applications may be made between R_2 and R_5.

A system developed at Iowa State University analyzes pod infection by *Phomopsis* spp. to predict the need for fungicide application. The system depends on field scouting and laboratory analysis of pods sampled at growth stage R_7.

A predictive system for the management of stem canker in the southern United States has been developed and incorporated into the Auburn University model. It uses information on cultivar susceptibility, weather, and time of infection to predict disease severity and losses at harvest uses. It also adjusts predictions of yield and crop value if a systemic fungicide (benomyl) is applied during predicted infection periods.

Both systemic and protective fungicides have been registered for use on soybeans. Both types have increased yields or improved seed quality. However, systemic fungicides, such as benomyl and thiophanate-methyl, have performed better than protective fungicides, such as chlorothalonil and mancozeb. Numerous researchers have found that correct applications with minimal drift losses and thorough canopy penetration are critical to the success of foliar fungicide programs on soybeans. Both aerial and ground application can be used; aerial application is generally more practical. Poor performance in fungicide programs can be consistently attributed to poor application techniques or situations in which disease pressure is light.

Soil-applied fungicides are effective for controlling Phytophthora root rot, Pythium root rot, and Rhizoctonia root rot. Effective management programs using soil-applied fungicides (e.g., metalaxyl) to control Phytophthora root rot also require tolerant cultivars and other cultural controls. Soil fungicides are generally applied in-furrow or in bands over the row.

Nematicides

Nematicides are used to control soybean cyst, root-knot, root-lesion, lance, dagger, stem, and reniform nematodes. For economic reasons, they are generally used only when adapted resistant cultivars are unavailable or when cultural controls such as rotation are ineffective. Most nematicides available today are granular insecticide-nematicide materials, such as aldicarb, carbofuran, and ethoprop. They are generally applied at planting, by either in-furrow or row-banded application. Fumigant nematicides, such as D-D (a mixture of 1,3-dichloropropene and 1,2-dichloropropane) or mixtures of D-D and chloropicrin, are applied in-row or broadcast prior to planting.

Insecticides

Insecticides are generally ineffective in controlling aphid-vectored viruses and are rarely used to control thrips-transmitted viruses, such as tobacco ringspot virus, or beetle-transmitted viruses, such as bean pod mottle virus. They might be considered to reduce fungal infections that follow pod feeding by insects such as the bean leaf beetle, grasshoppers, or stinkbugs. Decisions about the use of insecticides to reduce insect damage should take into account the added damage from opportunistic fungi. This is particularly true with stinkbugs that transmit yeast spot and the bean leaf beetle, whose feeding causes pods injuries through which several seed-infecting fungi can gain to entry.

Biological Control

Biological control agents, either applied to the seed or soil or stimulated by the addition of amendments, are presently the focus of a great deal of research. Nematode populations can be reduced by soil-inhabiting parasitic fungi, and mechanisms to stimulate populations of these fungi and thus achieve long-term biological control appear to be available. At present, no commercial products are registered or developed for biological control of soybean diseases. Treatment of seeds with rhizosphere-competent bacteria appears to have promise for the control of Rhizoctonia and Phytophthora root rots. Although commercial products are not yet available and the degree of control is not yet equal to that of alternative methods, it is likely that products for biological control will soon be developed, particularly as the effects of deleterious exopathogens in the rhizosphere are better understood.

Selected References

Backman, P. A., Crawford, M. A. and Hammond, J. M. 1984. Comparison of meteorological and standardized timings of fungicide applications for soybean disease control. Plant Dis. 68:44-46.

Jacobsen, B. J. 1979. Effect of foliar fungicides on soybean yield and seed quality. Proc. Soybean Seed Res. Conf., 9th, pp. 49-55.

Jacobsen, B. J. 1986. Calibration and use of aircraft for applying fungicides. Pages 39-44 in: Methods for Evaluating Pesticides for Control of Plant Pathogens. K. D. Hickey, ed. American Phytopathological Society, St. Paul, MN.

Jacobsen, B. J., Shurtleff, M. C., Kirby, H. W., and Melton, T. A. 1987. Condensed plant disease management guide for field crops. Univ. Ill.

Table 15. Illinois Checklist Used to Determine Whether Foliar Fungicide Should Be Applied to Soybeans[a]

Risk Factor	Point Value if Answer Is Yes[b]	
Rainfall, dew, and humidity up to early bloom and pod set are:		
Below normal	0	
Normal	2	
Above normal	4	______
Soybeans were grown in the field last year	2 to 3	______
Chisel plow, disk, or no-till was used	1	______
Pycnidia (black specks) are visible on fallen petioles[c] and Septoria brown spot is obvious on the lower leaves	2	______
Early-maturing variety (not full-season)	1 to 2	______
Soybeans are to be used or sold for seed	6	______
Yield potential is better than 35 bushels per acre	2	______
Seed quality at planting time is less than 85% germination in a warm test	1	______
Other conditions that favor disease development (weather forecast with a 30-day period of greater than normal rainfall and a field history of disease)	1 to 3	______
Total		______

[a] Source: Jacobsen et al, 1987.
[b] If the total point value is 15 or more, application will probably mean increased yields and higher seed quality.
[c] Only brown, fallen petioles should be assayed, and more than two thirds to three fourths of these petioles should show pycnidia.

Table 16. Kentucky Point System for Use of Foliar Fungicides for Soybean Seed Production[a]

	Points[b]	
I. Cropping history		
In soybeans the previous 2 or more years	3	______
In soybeans the previous year	2	______
Not in soybeans last year	0	______
II. Cultivar selection		
Early-season cultivar	3	______
Mid-season cultivar	2	______
Late-season cultivar—foliar fungicides should not be used on late-season cultivars		
III. Planting date		
Prior to May 20	3	______
Between May 20 and June 20	2	______
After June 20	0	______
IV. Rainfall—based on existing rainfall and rainfall predictions during seed development and maturation between R_2 and R_7 stages of growth		
Rainfall below normal	0	______
Rainfall near normal (±1.25 cm)	2	______
Rainfall above normal	4	______
Total		______

[a] Source: TeKrony et al, 1985.
[b] 11 points or more = fungicide should be applied; 9 or 10 points = fungicide may be beneficial; and 8 points or fewer = fungicide should not be applied. Foliar fungicides are not recommended for use in soybean seed production when any of the following conditions exist: 1) yield potential is less than 20 bu/acre (1,344 kg/ha), 2) heavy weed pressure exists, and 3) a late-season cultivar is planted.

Coop. Ext. Serv., Circ. 1231. 10 pp.

Lifshitz, R., Simonson, F. M., Scher, J. W., Kloepper, J. W., Rorick-Semple, C., and Zaleska, I. 1986. Effect of rhizobacteria on the severity of Phytophthora root rot of soybean. Can. J. Plant Pathol. 8:102-106.

Liu, Z., and Sinclair, J. B. 1987. *Bacillus subtilis* as a potential biological control agent for Rhizoctonia root rot of soybeans. (Abstr.) Phytopathology 77:1687.

McGee, D. C. 1988. Evaluation of current predictive methods for control of Phomopsis seed decay of soybeans. Pages 22-25 in: Soybean Diseases of the North Central Region. T. D. Wyllie and D. H. Scott, eds. American Phytopathological Society, St. Paul, MN.

Moore, W. F., ed. 1981. Soybean Cyst Nematode. Soybean Industry Resource Committee, Extension Service, U.S. Department of Agriculture, Washington, DC. 24 pp.

Morgan-Jones, G., and Rodríguez-Kábana, R. 1987. Fungal biocontrol for the management of nematodes. Pages 94-99 in: Vistas on Nematology. J. A. Veech and D. W. Dickson, eds. Society of Nematologists, Hyattsville, MD.

Schmitthenner, A. F. 1988. Phytophthora rot of soybean. Pages 71-80 in: Soybean Diseases of the North Central Region. T. D. Wyllie and D. H. Scott, eds. American Phytopathological Society, ST. Paul, MN.

Smith, E. F., and Backman, P. A. 1988. Soybean stem canker: An overview. Pages 47-55 in: Soybean Diseases of the North Central Region. T. D. Wyllie and D. H. Scott, eds. American Phytopathological Society, St. Paul, MN.

TeKrony, D. M., Stuckey, R. E., Egli, D. B., and Tomes, L. 1985. Effectiveness of a point system for scheduling foliar fungicides in soybean seed fields. Plant Dis. 69:962-965.

Cultural Controls

Cultural practices such as maintaining adequate availability of plant nutrients and proper soil pH, supplying adequate but not excessive water, weed control, avoiding excessive plant density, and planting high-quality seed in a favorable seedbed are effective in reducing damage from many diseases, because these practices reduce plant stress. Healthy, vigorous plants suffer less yield loss from diseases than plants already under stress. Table 14 lists cultural practices for controlling particular diseases.

Water management is particularly critical in root diseases. Saturated soils create favorable conditions for infection by *Pythium* and *Phytophthora*, and soil water deficits contribute significantly to losses caused by nematodes, charcoal rot, and brown stem rot. Excessively dry seedbeds can contribute to seed decay by *Aspergillus* and *Penicillium* spp. These fungi can decay seeds with moisture contents far less than the 50% moisture required for germination. Irrigation water management can be important in reducing diseases such as pod and stem blight, anthracnose, bacterial blight, Septoria brown spot, and Sclerotinia white mold. Furrow irrigation reduces pod and seed infections in many areas, whereas excessive sprinkler irrigation during pod fill and pod maturation increases them. In areas where Sclerotinia white mold is a threat, avoidance of irrigation during flowering helps reduce floral infection. Irrigation to maintain adequate moisture can reduce damage from charcoal rot and brown stem rot. In the control of brown stem rot, adequate moisture is particularly important during growth stages R_6–R_8.

Optimal soil pH (pH 6.2–7.0) is important for root nodulation. Acid soils accentuate the damage from Sclerotium blight. Suboptimal soil pH, which reduces nutrient availability, increases the damage from a number of diseases.

Optimal fertility, particularly adequate potash and phosphorus nutrition, is important in the management of bacterial blight, bacterial pustule, charcoal rot, pod and stem blight, Fusarium blight, Rhizoctonia root rot, and the soybean cyst nematode.

Excessive seeding rates and plant densities contribute to increased damage from charcoal rot, Sclerotium blight, and diseases caused by *Rhizoctonia* and *Pythium*. Therefore, seeding rates should be adjusted for soil fertility, soil moisture, seed quality, cultural factors, and the growth habit of the cultivar.

Cultivation can also influence disease development. Damage from Rhizoctonia root rot and Fusarium root rot can be minimized by cultivating soil against the stem to encourage the development of adventitious roots. This practice increases the damage from Sclerotium blight, however, and should be avoided if this disease is present. Cultivation when plants are wet spreads bacterial blight and bacterial pustule and is likely to spread other bacterial diseases.

Weeds growing in and around production fields can serve as hosts of fungal, bacterial, viral, and nematode pathogens of soybeans. Weed control in soybeans and rotation crops can be important in controlling the soybean cyst and root-knot nematodes. Uncultivated perennial plants growing around fields are often hosts of viruses, such as tobacco ringspot, bean pod mottle, soybean mosaic, and bean yellow mosaic viruses.

The selection of herbicides for weed control can influence disease development or control by other pesticides. For example, chloramben can increase the severity of Thielaviopsis root rot, and other herbicides increase the severity of Rhizoctonia and Phytophthora root rots. The herbicide alachlor reduces the effectiveness of nematicides in the control of the soybean cyst nematode. Weeds growing in soybean fields compete for space, water, and nutrients, thereby increasing the likelihood of stress and stress-related diseases.

Row spacing also affects soybean diseases. Narrow row widths have been reported to increase the severity of downy mildew, Rhizoctonia aerial blight, and Sclerotinia white mold but have little effect on Septoria brown spot. Where soybeans are planted early, the rapid canopy closure in narrow rows may lower soil temperatures, reduce the severity of charcoal rot, and suppress damaging weed populations.

Soybean growth habit is another factor in cultural control. Septoria brown spot and pod and stem blight are more severe, causing greater yield losses, in dwarf or semidwarf determinant soybeans than in nondwarf cultivars.

Seedbed conditions that favor rapid seed germination and emergence minimize damage from pre- and postemergence damping-off. Soybean seeds germinate when soil temperatures are above 12–14°C and seed moistures exceed 50%. Where these conditions are not met, the use of a seed treatment fungicide is suggested, even if high-quality seed is used. If conditions are favorable for germination and emergence, the need for seed treatment is dependent on seed quality.

The planting date can also influence diseases other than those that affect seedlings. Early planting tends to minimize damage from the soybean cyst nematode, charcoal rot, and stem canker. It minimizes the likelihood that soybeans will be in the critical growth stages for infection with stem canker during wet periods. However, early planting can accentuate damage from pod and stem blight, particularly if conditions in middle to late summer favor this disease.

Selected References

Balducchi, A. J., and McGee, D. C. 1987. Environmental factors influencing infection of soybean seeds by *Phomopsis* and *Diaporthe* species during seed maturation. Plant Dis. 71:209-212.

Bostian, A. L., Schmitt, D. P., and Barker, K. R. 1986. Population changes of *Heterodera glycines* and soybean yields resulting from soil treatment with alachlor, fenamiphos, and ethoprop. J. Nematol. 18:458-463.

Bowman, J. E., Hartman, G. L., McClary, R. D., Sinclair, J. B., Hummel, J. W., and Wax, L. M. 1986. Effects of weed control and row spacing in conventional tillage, reduced tillage, and nontillage on soybean seed quality. Plant Dis. 70:673-676.

Cerkauskas, R. F., Dhingra, O. D., Sinclair, J. B., and Asmus, G. 1983. *Amaranthus spinosus, Leonotis nepetaefolia*, and *Leonurus sibiricus*: New hosts of *Phomopsis* spp. in Brazil. Plant Dis. 67:821-824.

Dirks, V. A., Anderson, T. R., and Botton, E. F. 1980. Effect of fertilizer and drain location on incidence of Phytophthora root rot of soybean. Can. J. Plant Pathol. 2:179-183.

Duncan, D. R., and Paxton, J. D. 1981. Trifluralin enhancement of Phytophthora root rot of soybean. Plant Dis. 65:435-436.

Table 17. Guidelines for Seed Indexing and Need for Fungicide Seed Treatment, Used by Illinois Crop Improvement Association[a]

Seed Quality Rating	Seed Lot Performance by Test			Fungicide Recommendation
	Warm Germination (%)	Cold Germination (%)	Diseased Seeds (%)	
High quality	Greater than 85	Greater than 70	Less than 10	Seeds do not need treatment unless planted where germination may be delayed or where planting rates are low
Average to low quality	75 to 85	60 to 69	10 to 20	Treat
Poor quality	Less than 75	Less than 60	Greater than 20	Do not use or reclean; treat and retest samples; use higher seedling rate and plant in soils >55° F

[a] Adapted from Jacobsen, 1985. Guidelines developed by the Illinois Crop Improvement Association, Urbana.

Ferriss, R. S. 1988. Seedling establishment—An epidemiological perspective. Pages 7-13 in: Soybean Diseases of the North Central Region. T. D. Wyllie and D. H. Scott, eds. American Phytopathological Society, St. Paul, MN.

Grau, C. R., and Radke, V. L. 1984. Effects of cultivars and cultural practices on Sclerotinia stem rot of soybean. Plant Dis. 68:56-58.

Helbig, J. B., and Carroll, R. B. 1984. Dicotyledonous weeds as a source of *Fusarium oxysporum* pathogenic on soybean. Plant Dis. 68:694-696.

Hepperly, P. R., Kirkpatrick, B. L., and Sinclair, J. B. 1980. *Abutilon theophrasti*: Wild host for three fungal parasites of soybean. Phytopathology 70:307-310.

Jeffers, D. L., Schmitthenner, A. F., and Kroetz, M. E. 1982. Potassium fertilization effects on Phomopsis seed infection, seed quality, and yield of soybeans. Agron. J. 74:886-890.

Mengistu, A., Tachibana, H., Epstein, A. H., Bidne, K. G., and Hatfield, J. D. 1987. Use of leaf temperature to measure the effect of brown stem rot and soil moisture stress and its relation to yield of soybeans. Plant Dis. 71:632-634.

Schmitt, D. P., and Nelson, L. A. Interactions of nematicides with other pesticides. Pages 455-460 in: Vistas on Nematology. J. A. Veech and D. W. Dickson, eds. Society of Nematologists, Hyattsville, MD.

Wyllie, T. D. 1988. Charcoal rot of soybeans—Current status. Pages 106-113 in: Soybean Diseases of the North Central Region. T. D. Wyllie and D. H. Scott, eds. American Phytopathological Society, St. Paul, MN.

Seed Quality Assessment

Several tests can be used to assess soybean seed quality. In most situations, accurate assessment requires the use of several testing methods. In temperate environments, indexing seed lots by means of a standard warm-germination test, a cold-soil test, and an assessment of seedborne *Phomopsis* sp. provides an excellent evaluation of seed quality. Guidelines developed and used by the Illinois Crop Improvement Association (Urbana) to index seed lots are given in Table 17. The standard warm-germination test is conducted on moist cellulose pads (Fig. 78) at 23–25°C for 7 days. The percentage of seeds infected with *Phomopsis* sp. is noted, and the percentage of germinated seeds is determined at the same time. The cold-soil germination test is performed with a 1:1 mix of field soil and sand wetted to field moisture capacity. Seeded containers are kept at 10°C for 7 days and then at 23–25°C for 5 days. Only strong, vigorous seedlings are counted. Some laboratories use a mixture of sterile sand and vermiculite for cold-vigor tests. However, mixtures of field soil and sand give better predictions of field performance.

Under more tropical conditions, an accelerated aging test may be useful, in addition to the standard germination and fungal infection tests. In the accelerated aging test, seeds are placed in wire-mesh baskets at 100% relative humidity and 41°C for 3 days and then are placed on cellulose pads or rolled towels at 23–25°C for 4 days.

Physical or mechanical seed damage can be assessed by tetrazolium chloride tests or by soaking seeds in a 1% sodium hypochlorite solution for 1 hr. The tetrazolium test identifies cracks and dead or damaged areas in soybean seed tissues. The bleach soak identifies seeds with cracked or damaged seed coats. Mechanical damage is a serious problem in harvesting or handling seeds with moisture contents of less than 10%.

Seed cleaning is an important step in producing high-quality seed lots. Gravity tables or spiral cleaners remove a large percentage of moldy seed; small, lightweight seed; fungal sclerotia; and soil peds. Removal of soil peds can be important in reducing the spread of the soybean cyst nematode, whose cysts can be found in them.

Selected References

Clark, B. E., McDonald, M. B., Jr., and Joo, K., eds. 1983. Seed Vigor Testing Handbook. Association of Official Seed Analysts. Madison, WI. 173 pp.

Delouche, J. C., and Baskin, C. C. 1973. Accelerated aging techniques for predicting the relative storability of seedlots. Seed Sci. Technol. 1:427-452.

Franca Neto, J. B., and Henning, A. A. 1984. Physiological and pathological seed quality of soybeans. (In Portuguese) EMBRAPA-CNPSoja, Circ. Tec. 9. 39 pp.

Jacobsen, B. J. 1985. Soybean seed quality and use of fungicide seed treatments. Ill. Seed News 15(10):2-4.

Matthews, S. 1981. Evaluation of techniques for germination and vigor studies. Seed Sci. Technol. 9:543-551.

Disease Resistance

The use of disease-resistant cultivars is the most economical and efficient method of disease control if they yield competitively. Where they yield less than susceptible cultivars, resistant cultivars can be used effectively in rotations on a prescription basis. Prescription varieties have been effective in control programs for the soybean cyst nematode and for brown stem rot. Resistance is often not available in adapted cultivars or is race-specific. Race-specific resistance is often a problem in cultivars used for controlling the soybean cyst nematode, Phytophthora root rot, soybean rust, frogeye leaf spot, and downy mildew, since numerous races of the causal pathogens are known. In these cases it is important to know the predominant race present. Diseases controlled by resistant varieties are listed in Table 14. More detailed information is given in the descriptions of individual diseases.

Selected References

Hartwig, E. E., and Edwards, C. J. 1986. Evaluation of Soybean Germplasm Maturity Groups V Through X. U.S. Department of Agriculture, Agricultural Research Service, Stoneville, MS. 156 pp.

Nelson, R. L., Amdor, P. J., Orf, J. H., Lambert, J. W., Cavins, J. F., Kleiman, R., Laviolette, F. A., and Athow, K. L. 1987. Evaluation of the USDA soybean germplasm collection Maturity Groups 000–IV (PI 273.483 to PI 427.107). U.S. Dep. Agric., Agric. Res. Serv., Tech. Bull. 1718. 267 pp.

Shibles, R., ed. 1985. Disease and insect resistance. Pages 407-438 in: Proc. World Soybean Res. Conf. III. Westview Press, Boulder, CO.

Tachibana, H. 1988. Use and management of resistance for control of brown stem rot of soybeans. Pages 102-105 in: Soybean Diseases of the North Central Region. T. D. Wyllie and D. H. Scott, eds. American Phytopathological Society, St. Paul, MN.

Tisselli, O., Sinclair, J. B., and Hymowitz, T. 1980. Sources of Resistance to Selected Fungal, Bacterial, Viral, and Nematode Diseases of Soybean. INTSOY Ser. 18. College of Agriculture, University of Illinois at Urbana-Champaign. 134 p.

Integrated Pest Management

Management of soybean diseases can rarely be accomplished in the long term by only one control method; disease management must be placed within the context of agronomic practices, weed or insect control, and economics. Integrated pest management, which utilizes economic injury levels, economic thresholds, scouting, record keeping, and pest mapping, allows for a planned, economically sound approach to crop management. Further, it is becoming increasingly important to identify pathogens to the race level. Integrated pest management programs must be adapted to local situations, and several have been developed, including some that use computer software.

One such program is the Auburn University Soybean Integrated Pest Management Model (AUSIMM), developed to predict the profitability of soybean management practices in the southeastern United States. The model begins with a ranking of about 120 cultivars according to maturity group, relative productivity, and susceptibility to nematodes and diseases. Potential yields of optimal cultivars are calculated, given the location, soil type, planting date, rotation, and pest histories of individual fields. Submodels are utilized later for management of nematodes, plant disease, and insect pests. These integrate data on soybean plant growth, population dynamics of major pests, pesticide efficacy, weather, effects of weather on pest populations and pesticide efficacy, and crop loss functions. The effectiveness of the control practices is compared to their cost. Submodels are linked by a core program, which estimates potential yield and crop value through the season, accumulating changes in potential yield, crop value, and management costs.

Selected References

Frisbie, R. E., and Adkisson, P. L., eds. 1985. Integrated Pest Management on Major Agricultural Systems. Texas A&M University, College Station. 743 pp.

Herbert, D. A., Backman, P. A., Rodríguez-Kábana, R., and Mack, T. P. 1988. Disease, insect, and nematode submodels of AUSIMM: The Auburn University Soybean Integrated Pest Management Model. (Abstr.) Proc. South. Soybean Dis. Work. 15:71.

(Prepared by B. J. Jacobsen and P. A. Backman)

Index

Water stress, 88. *See also* Drought; Flooding
Watermelon mosaic virus 2, 51
"Weather damage," 81
Weed control, for disease control, 95, 98
Whetzelinia sclerotiorum, 47
White clover mosaic virus, 51
White mold. *See* Sclerotinia white mold;
  Sclerotium blight
Whitefly, 61, 62
Wildfire, 7–8, 95
Wisconsin pea streak virus, 51
Witches'-broom, 9, 10

Xanthomonas, 4
  campestris, 85
    pv. *campestris*, 4
    pv. *cannabis*, 4, 9
    pv. *glycines*, 4, 6, 8, 74
    pv. *phaseoli*, 4, 6
  phaseoli var. *sojensis*, 6
Xiphinema (dagger nematodes), 63, 64, 97
  americanum, 59, 63, 64, 72
  diversicaudatum, 64, 72

Yeast spot, 79, 95

Yellow dwarf disease. *See Heterodera
  glycines*
Yellow leaf spot of unknown cause, reported
  in Thailand, 86
Yellow mosaic virus, 51, 95

Zea mays, 94
Zinc, as nutrient, 89, 90
  nematode injury similar to deficiency of,
  64
Zinnia elegans, 58, 62